Workbook to Accompany Welding and Metal Fabrication

Larry Jeffus

Cengage

Australia • Brazil • Canada • Mexico • Singapore • United Kingdom • United States

T0336455

Workbook to Accompany Welding and Metal Fabrication
Larry Jeffus

Vice President, Editorial: Dave Garza

Director of Learning Solutions: Sandy Clark

Executive Editor: Dave Boelio

Managing Editor: Larry Main

Senior Product Manager: Sharon Chambliss

Editorial Assistant: Jillian Borden

Vice President, Marketing: Jennifer Baker

Executive Marketing Manager:
Deborah S. Yarnell

Marketing Specialist: Mark Pierro

Production Director: Wendy Troeger

Production Manager: Mark Bernard

Senior Content Project Manager: Cheri Plasse

Senior Art Director: Joy Kocsis

Technology Project Manager:
Christopher Catalina

For product information and technology assistance, contact us at
**Cengage Customer & Sales Support, 1-800-354-9706
or support.cengage.com.**

For permission to use material from this text or product, submit all requests online at **www.copyright.com.**

Library of Congress Control Number: 2010931182

ISBN-13: 978-1-4180-1375-2
ISBN-10: 1-4180-1375-7

Cengage
200 Pier 4 Boulevard
Boston, MA 02210
USA

Cengage is a leading provider of customized learning solutions with employees residing in nearly 40 different countries and sales in more than 125 countries around the world. Find your local representative at: **www.cengage.com.**

To learn more about Cengage platforms and services, register or access your online learning solution, or purchase materials for your course, visit **www.cengage.com.**

Notice to the Reader
Publisher does not warrant or guarantee any of the products described herein or perform any independent analysis in connection with any of the product information contained herein. Publisher does not assume, and expressly disclaims, any obligation to obtain and include information other than that provided to it by the manufacturer. The reader is expressly warned to consider and adopt all safety precautions that might be indicated by the activities described herein and to avoid all potential hazards. By following the instructions contained herein, the reader willingly assumes all risks in connection with such instructions. The publisher makes no representations or warranties of any kind, including but not limited to, the warranties of fitness for particular purpose or merchantability, nor are any such representations implied with respect to the material set forth herein, and the publisher takes no responsibility with respect to such material. The publisher shall not be liable for any special, consequential, or exemplary damages resulting, in whole or part, from the readers' use of, or reliance upon, this material.

Printed at CLDPC, USA, 12-23

CONTENTS

INTRODUCTION

There are many job opportunities for welders who have demonstrated their ability to both fit up a weldment and then make acceptable welds to complete the project. Your job may require you to make welds in various positions with any of the major welding processes. The experiments, practices, and projects in *Welding and Metal Fabrication* are designed to help you develop your welding and fitting skills. Each of the welding and cutting processes covered in the textbook follow a logical progression from the basics to more technically challenging projects.

Many jobs require more than just your ability to fit up and weld; they may also expect you to have a basic understanding of the science of welding. Your technical knowledge can be developed in the classroom, and your welding and fitting skills can be developed in the welding lab. The application of your technical knowledge can be refined by using this workbook.

This workbook contains a number of exercises and quizzes that are designed to help you identify areas of welding technology you need to better understand. Often as we read and study new material, we think we clearly understand everything that was presented. Once we begin answering questions on quizzes or identifying material in the figure exercises, we can identify areas that may not have been as clearly understood as we had thought. When you find an item you are not sure you know the answer to, leave that answer blank. The questions you leave blank can be more important to you than the ones you get right because they help you identify areas where you need additional study. Once you have identified the areas, you and your instructor can work together to help you understand that material. After all, the grade you receive in class on a quiz may let you pass the course, but the understanding you have of the material will help you become a successful career welder.

QUIZZES

CHAPTER 1

Introduction to Welding

Quiz

Name: _____

Date: _____

Class: _____

Instructor: _____

Grade: _____

Instructions: Carefully read Chapter 1 in the text and answer the following questions:

MULTIPLE-CHOICE

1. What do these items all have in common: trains, furniture, recreational vehicles, and buildings?
   a. they are all related to transportation
   b. they are all made of metal
   c. they can all be manufactured by welding
   d. they are all things you might find in someone's home

2. Which of the following items would require some type of welding?
   a. pencil
   b. computer
   c. spiral notebook
   d. textbook

3. In metal fabrication what step must be done before parts can be cut out?
   a. assemble
   b. weld
   c. finish
   d. lay out

4. Which welding process uses a 14-in. long flux-covered electrode?
   a. SMAW
   b. TB
   c. OSC
   d. PAC

5. Which two processes use a continuously fed wire?
   a. SMAW and TB
   b. FCAW and GMAW
   c. OFC and OAW
   d. OFW and LBC

6. Which welding processes use a shielding gas that flows from the gun nozzle to protect the weld metal?
   a. PRW and RW
   b. EASP and SMAW
   c. OFC and TB
   d. GTAW and GMAW

7. Which of the following is a thermal cutting process that can be used on almost any metal or alloy?
   a. PAC
   b. OFC
   c. OAC
   d. TB

8. What might be necessary if parts are not accurately assembled before welding?
   a. the part may have to be painted a different color
   b. more tack welds might be required
   c. the weld might have to be removed and the part repositioned
   d. nothing, this does not cause a problem

9. Who performs the actual welding?
   a. engineer
   b. welding salesman
   c. welding inspector
   d. welder

10. Inches can be converted to which SI unit?
    a. 1
    b. mm
    c. kPa
    d. c

SENTENCE COMPLETION

In the space provided write the answer that completes the statement.

1. The primary steps for fabrication include layout, cut out, _____, welding, and finishing.

2. The term _____ means the fusion or growing together of the grain structure of the materials being welded.

3. _____ is the joining together of the surface(s) of a material by the application of heat only, pressure only, or with heat and pressure together so that the surfaces fuse together.

4. An example of a welded object that would need to have a high standard of quality to be fit for service is _____.

5. An example of a welded object that would not need a high standard of quality to be fit for service is _____.

6. The advantages of _____ welding are that it is extremely fast, is economical, can produce long welds rapidly, needs very little postweld cleanup, and can weld thicknesses from thin-gauge sheet metal to heavy plate.

7. The advantages of _____ welding are that it produces high quality welds, is versatile, fast, and cost effective.

8. GMA and FCA welding are called _____ welding processes because the filler metal is automatically fed into the welding arc, and the welder manually moves the welding gun along the joint being welded.

9. The _____ forms the gaseous cloud during shielded metal arc welding.

10. The advantages of _____welding are that it is the cleanest of all of the manual welding processes; it can be used to make extremely high-quality welds in applications in which weld integrity is critical; and there are metal alloys that can be joined only with this welding process.

11. The most common fuel gas used with oxygen for cutting, welding, and brazing is _____.

12. Oxyfuel gas cutting, plasma arc cutting, air carbon arc cutting, and laser beam cutting are commonly used _____ cutting processes.

13. Oxyfuel gas cutting can be used on steel from _____ thick to _____ thick, depending on the capacity of the torch and tip being used.

14. An advantage of the _____ cutting process is that almost any metal or alloy can be cut, and the high travel speeds and very low heat input help to reduce or eliminate part distortion.

15. Welding inspectors are often required to hold a special certification such as the one supervised by the American Welding Society known as _____ _____.

MATCHING

In the space provided to the left, write the letter from Column B that describes the abbreviation in Column A.

Column A	Column B
_____ 1. AWS	a. gas tungsten arc welding
_____ 2. FCAW	b. oxyfuel gas cutting
_____ 3. FOW	c. shielded metal arc welding
_____ 4. GMAW	d. American Welding Society
_____ 5. GTAW	e. forge welding
_____ 6. OF	f. torch or oxyfuel brazing
_____ 7. TB	g. flux cored arc welding
_____ 8. OFC	h. gas metal arc welding
_____ 9. OFW	i. oxyfuel gas welding
_____ 10. SMAW	j. oxyfuel gas

CHAPTER 2
Welding Safety

Quiz

Name: _____
Date: _____
Class: _____
Instructor: _____
Grade: _____

Instructions: Carefully read Chapter 2 in the text and answer the following questions:

MULTIPLE-CHOICE

1. Reading, learning, and following all safety rules, regulations, and procedures is the responsibility of
   a. your teacher
   b. your boss
   c. yourself
   d. MSDS

2. A burn where the surface of the skin is severely damaged, resulting in the formation of blisters and possible breaks in the skin is classified as a
   a. first-degree burn
   b. second-degree burn
   c. third-degree burn
   d. burn

3. A burn where the surface of the skin and possibly the tissue below the skin appear white or charred, there may be cracks or breaks in the skin, and there is little pain because nerve endings have been destroyed is classified as a
   a. first-degree burn
   b. second-degree burn
   c. third-degree burn
   d. burn

7

4. A burn where the surface of the skin is reddish in color, tender, and painful but does not involve any broken skin is classified as a
   a. first-degree burn
   b. second-degree burn
   c. third-degree burn
   d. burn

5. Protecting the welder against hot sparks, metal, and slag is the purpose of
   a. proper ventilation
   b. welding curtains
   c. double insulation
   d. protective clothing

6. The best way to avoid inhaling dangerous fumes when welding is to provide proper
   a. welding curtains
   b. ventilation
   c. protective clothing
   d. goggles

7. When lifting a heavy object, your _____ should be used to lift, not your _____.
   a. back, legs
   b. arms, back
   c. legs, back
   d. arms, legs

8. Falls from ladders are commonly caused by
   a. metal ladders
   b. dizziness
   c. wooden ladders
   d. improper use of ladders

9. How far must highly combustible materials like paint, fuels, and oil be kept from a welding or cutting area?
   a. 15 ft (4.75 m)
   b. 35 ft (10.7 m)
   c. 10 ft (3.05 m)
   d. 20 ft (6.1 m)

10. Most electric shocks in the welding industry occur from
    a. accidental contact with bare or poorly insulated conductors
    b. dull tools
    c. missing labels
    d. cylinder leaks

SENTENCE COMPLETION

In the space provided write the answer that completes the statement.

1. A welder can get specific safety information for tools and equipment from welding equipment _____ and their local suppliers.

2. If an accident occurs on a welding site, the consequences can be that the job site may be _____ temporarily or permanently.

3. Burns can be caused by _____ light and contact with hot welding material.

4. The first step in treating a _____-degree burn is to put the burned area under cold water (not iced) or apply cold water compresses.

5. If the welding cannot be moved away from other workers, the area should be screened off with _____ that will absorb the welding light.

6. Welders should wear _____ when welding to protect their face and eyes.

7. The two parts of the eye that can be burnt by ultraviolet light are the _____ _____ and the _____ of the eyes.

8. The lens of a welding helmet can be checked for cracks by _____ _____ it between your fingers.

9. Two types of ear protection are _____ and _____ _____.

10. All welding and cutting processes produce undesirable by-products such as harmful __ _____.

11. Welding on dirty or used metal can cause chemicals that are on the metal to become mixed with the welding fumes, a combination that can be extremely _____ _____.

12. Forced _____ is always required when welding on metals that contain zinc, lead, beryllium, cadmium, mercury, copper, austenitic manganese, or other materials that give off dangerous fumes.

13. All manufacturers of potentially hazardous materials must provide _____ _____ giving detailed information regarding possible hazards resulting from the use of their products.

14. Throwing hazardous waste material into the _____, pouring it on the _____, or dumping it down the _____ is illegal.

15. Ladders can become worn or damaged over time and should be _____ _____ each time they are used.

16. Accidental contact with _____ causes most electric shocks in the welding industry.

17. Welding circuits must be turned off when the workstation is left unattended to prevent _____.

18. For protection from electrical shock, the standard portable tool is built with either of two equally safe systems: _____ or _____ _____.

19. A tool with external grounding has a wire that runs from the housing through the power cord to a _____ on the power plug.

20. Using a power source with a voltage lower than the rating on the nameplate is harmful to the _____.

21. An extension cord may be overloaded if it feels more than slightly _____ _____ to a bare hand placed outside the insulation.

22. Always connect the cord of a portable electric power tool into the extension cord before the extension cord is connected to the _____.

23. _____ is a good choice of fabric for protective clothing because it is flexible, resists burning, and is readily available.

24. If a leak cannot be stopped by closing the valve on a cylinder, it should be moved to a _____.

25. The location of fire extinguishers should be marked near the _____ _____ so they can be found even if a room is full of smoke.

26. A welder might use _____ on the job for assembly and disassembly of parts for welding as well as to perform routine equipment maintenance.

27. Hand tools should be kept _____ to make work easier, improve the accuracy of the work, and make them safer to use.

28. If a slight _____ is felt while using a power tool, you should stop and have the tool checked by an electrical technician.

29. A grinding stone can be checked for cracks by _____ the stone in four places and listening for a sharp ring, which indicates it is good.

30. When possible, _____ the workpiece before drilling to prevent the drill bit from moving across the surface being drilled as the drilling begins.

31. When using a hoist or crane, the load should be kept as close to the _____ _____ as possible while it is being moved.

MATCHING

In the space provided to the left, write the letter from Column B that best answers the question in Column A.

Column A	Column B
_____ 1. In what type of burn is there not much pain initially because nerve endings have been destroyed?	a. 35 feet (10.7 m)
_____ 2. What color can welding shops be painted to reduce the danger from reflected light?	b. 20 feet (6.1 m)
_____ 3. What part of your body should be used to lift a heavy object?	c. third-degree burn
_____ 4. How far must highly combustible materials be kept from a welding or cutting area?	d. external grounding
_____ 5. What type of light can cause a burn but cannot typically be seen or felt?	e. green
_____ 6. How far must a storage area for oxygen and fuel gas cylinders be from the welding area?	f. black
_____ 7. In what type of burn is the skin reddish in color with no broken skin?	g. electrical wires
_____ 8. What electrical safety system can portable tools have?	h. first-degree burn
_____ 9. What color of hose should be used only for oxygen?	i. legs
_____ 10. What should you never use ladders around?	j. ultraviolet light

30. **When possible,** _____ the workpiece before drilling to prevent the drill bit from moving across the surface being drilled as the drilling begins.

31. **When using a hoist or crane,** the load should be kept as close to the _____ as possible while it is being moved.

MATCHING

In the space provided to the left, write the letter from Column B that best answers the question in Column A.

Column A	Column B
1. In what type of burn is there not much pain initially, because nerve endings have been destroyed?	a. 35 feet (10.7 m)
2. What color can welding shops be painted to reduce the glare from reflected light?	b. 20 feet (6.1 m)
3. What area of your body should be used to lift a heavy object?	c. third-degree burn
4. How far must highly combustible materials be kept from a welding or cutting area?	d. external grounding
5. What type of light can cause a sunburn that cannot usually be seen or felt?	e. green
6. How far must a storage area for oxygen and fuel gas cylinders be from the welding area?	f. blue
7. In what type of burn is the skin reddish in color with no broken skin?	g. electrical wires
8. What electrical safety system can portable tools have?	h. first degree burn
9. What color of hose should be used only for oxygen?	i. legs
10. What should you never use ladders around?	j. ultraviolet light

CHAPTER 3

Shop Math

Quiz

Name: _____

Date: _____

Class: _____

Instructor: _____

Grade: _____

Instructions: Carefully read Chapter 3 in the text and answer the following questions:

MULTIPLE-CHOICE

1. Before fractions can be added or subtracted, they must
   a. have the decimal points lined up vertically
   b. be reduced
   c. have a common denominator
   d. be rounded

2. To convert a fraction to a decimal, divide the _____ by the _____.
   a. numerator, denominator
   b. denominator, numerator
   c. fraction, decimal
   d. decimal, fraction

3. Find the total length of two pieces of pipe when one is 6.75 ft long and the other is 4.5 ft long.
   a. 10.25 ft
   b. 2.25 ft
   c. 11.5 ft
   d. 11.25 ft

4. Bar stock comes in 10-ft and 20-ft lengths. If four 6-ft long pieces of bar stock will be cut, how many total feet of bar stock will have to be purchased?
   a. 20 ft
   b. 30 ft
   c. 40 ft
   d. 50 ft

5. What is the center of a length of pipe that is 12 ft 8 in. long?
   a. 6 ft 4 in.
   b. 4 ft 6 in.
   c. 3 ft 4 in.
   d. 6 ft 6 in.

6. If you need to cut a 3 ft 6 in. piece and a 7 ft 8 in. piece of steel bar, which of the following scrap pieces can be used with the least scrap left over after the cuts?
   a. 10 ft 6 in.
   b. 11 ft
   c. 12 ft 2 in.
   d. 13 ft 1 in.

7. How much scrap pipe will you have if you cut out a 7 ft 10 in. length from a 20-ft length of pipe?
   a. 13 ft 2 in.
   b. 27 ft 10 in.
   c. 13 ft
   d. 12 ft 2 in.

8. A welder uses 2.675 cu ft of acetylene gas to cut one angle iron. How much acetylene gas would be needed to cut 30 angle irons?
   a. 11.21 cu ft
   b. 80.25 cu ft
   c. 32.675 cu ft
   d. 82.5 cu ft

9. Round 52.126 to the second decimal place.
   a. 52.1
   b. 52.2
   c. 52.126
   d. 52.13

10. An acceptable dimension for an 8-in. long part with a tolerance of $\pm 1/8$ in. would be a minimum of _____ inches and a maximum of _____ inches.
    a. 7 1/2 in., 8 1/2 in.
    b. 8 in., 8 1/8 in.
    c. 7 7/8 in., 8 1/8 in.
    d. 7 3/4 in., 8 1/4 in.

SENTENCE COMPLETION

In the space provided write the answer that completes the statement.

1. The most common use of math in welding fabrication is for _____
  _____.

2. You cannot add feet to inches without first converting the feet to _____
  _____.

3. Examples of whole numbers are _____.

4. A _____ is a number that uses a decimal point to
  denote a unit that is smaller than 1.

5. The most common types of _____ used in welding
  fabrication are linear dimensions, angular dimensions, weight, and time.

6. A _____ is two or more numbers that are used to
  express a unit smaller than one.

7. Examples of fractions are _____.

8. The _____ is the bottom number of a fraction, and the
  _____ is the top number.

9. An _____ is a mathematical statement in which both
  sides are equal to each other, for example, 2X = 1Y.

10. In the sequence of mathematical operations—1st perform all operations within
  _____ .

11. In the sequence of mathematical operations—2nd resolve any _____
  _____ .

12. In the sequence of mathematical operations—3rd do all _____
  _____ and _____ working from left to right.

13. In the sequence of mathematical operations—4th do all _____
  _____ and _____ working from left to right.

14. _____ is the same thing as adding the same number
  to itself over and over.

15. _____ is the process of subtracting the same number
  over and over again.

16. Decimal fractions are added and subtracted just like whole numbers as long as the
  decimal points are kept in a _____ line.

17. When multiplying decimal fractions, you need to ensure that the _____ _____ is placed in the correct position in the answer.

18. The decimal point in the answer of a division problem is placed before any _____ _____ are done.

19. When rounding off a number, look at the number to the right of the last significant place to be used; if this number is less than _____, drop it and leave the remaining number unchanged.

20. Before fractions can be added or subtracted, they must have a common _____ _____.

21. To convert a denominator, multiply both the numerator and the denominator of the fraction by the same _____.

22. The normal way to reduce a fraction is to find the _____ number that can be divided into both the denominator and numerator.

23. To convert a fraction to a decimal, divide the _____ (top number in the fraction) by the _____ (bottom number in the fraction).

24. Most drawings usually state a dimensioning _____, the amount by which a part can be larger or smaller than the stated dimensions and still be acceptable.

25. To convert a decimal to a fraction, multiply the decimal by the _____ _____ of the fractional units desired; that is, for 8ths (1/8) use 8, for 4ths (1/4) use 4, and so on.

MATCHING

In the space provided to the left, write the letter from Column B that best completes the statement in Column A.

Column A	Column B
_____ 1. 7 is an example of a	a. fractions
_____ 2. wt = [(l″ × w″ × t″) ÷ 1728] × wt/ft is an example of a	b. mixed unit
_____ 3. 2X = 1Y is an example of an	c. rounding
_____ 4. 1/4, 1/2, 3/8, and 5/16 are examples of	d. sequence of mathematical operations
_____ 5. 5 3/4 is an example of a	e. tolerance
_____ 6. .75 is an example of a	f. whole number

_____ 7. The name of the term for the *one* in 1/2 is

_____ 8. The name of the term for the *two* in 1/2 is

_____ 9. Changing 2/4 to 1/2 is called

_____ 10. Before adding 2/16, 2/4, and 5/8, you must find the

_____ 11. When you change 13 in. to 1 ft 1 in. it is called

_____ 12. Changing 4.376 to 4.38 is an example of

_____ 13. Performing all operations within parentheses is the first step in the

_____ 14. ±1/16 in. is an example of a

_____ 15. The measurements of an object, such as 5 ft 7 in. or 2 ft 6 in. are called

g. denominator

h. dimensions

i. converting

j. decimal fraction

k. equation

l. common denominator

m. formula

n. numerator

o. reducing

WORD PROBLEMS

In the space provided write the answers to the following math word problems.

1. Round off the following numbers to the second decimal place: (A) 64.294, (B) 123.586, (C) 4.215. (A) _____ (B) _____
_____ (C) _____

2. Round off the following numbers to the third decimal place: (A) 15.2386, (B) 6.7425, (C) 12.6752. (A) _____
(B) _____ (C) _____

3. An acceptable dimension for a 12-in. long part with a tolerance of ±1/8 in. would be a minimum of _____ inches and a maximum of
_____ inches.

4. An acceptable dimension for a 10-in. long part with a tolerance of ±1/16 in. would be a minimum of _____ inches and a maximum of
_____ inches.

5. An acceptable dimension for a 6-in. long part with a tolerance of ±0.125 in. would be a minimum of _____ inches and a maximum of
_____ inches.

6. If a 12-foot long piece of pipe is cut out of a 20-ft length of pipe, how much scrap pipe would be left? _____

7. To determine the root area for a groove weld the following formula is used: Root area = root opening × plate thickness. Determine the root area for groove welds with the following root openings and plate thicknesses:

   Root Opening × Plate Thickness = Root Area

   (A) 1/8 in. × 1/2 in. = _____

   (B) 1/4 in. × 3/8 in. = _____

8. Find the cost of labor for a job where the total number of hours worked is 8 hours and the rate of pay is $15 per hour. _____

9. How many total feet of pipe would be needed if the following sections are needed: One 6 ft 8 in. length, two 4 ft 1 in. lengths, and one 5 ft 2 in. length? _____
_____

10. How much scrap pipe will be left if a 13-ft 6 in. piece is cut out of a 20-foot length of pipe? _____

11. Bar stock comes in 10-ft and 20-ft lengths. If three 8-ft long pieces of bar stock will be cut, what is the minimum length of bar stock that will have to be purchased?

   _____

12. What is the center of a length of pipe that is 10 ft 6 in. long? _____
_____

13. Find the total length of two pieces of pipe if one is 6.5 ft and the other is 5.75 ft

   _____

14. How many 2.25-ft pieces of angle iron can be cut out of a 9-ft long piece? _____
_____

15. If two pieces of bar stock are needed: A piece 4 ft 3 in. long and a piece 7 ft 8 in. long, how many total feet of bar stock are needed? _____

16. How many feet of scrap pipe would be left from a 10-ft long piece if a 5-ft 1 in. long piece is cut off? _____

17. What is the total length of two pieces of steel bar if a piece is 2 ft 8 in. long and a piece is 4 ft 6 in. long?_____

18. What is the total length of these four pieces of angle iron: 2 ft 5 in., 3 ft 6 in., 4 ft 7 in., and 1 ft 10 in.?_____

19. Using the formula: Area = length × width, find the area of the bed of a utility trailer that is 6 ft long and 4 ft 6 in. wide. _____

20. How many 1/2-in. spacers can be cut out of a 4-1/2 in. long piece of bar stock?

   _____

PRACTICAL MATH

Solve the following math problems.

Add the following whole numbers:

1. 84 + 25 = _____

2. 123 + 68 = _____

3. 259 + 13 + 37 = _____

4. 537 + 28 + 116 = _____

5. 2432 + 1067 + 319 = _____

6. 858 + 2046 + 1102 = _____

Subtract the following whole numbers:

7. 869 − 437 = _____

8. 698 − 366 = _____

9. 748 − 629 = _____

10. 537 − 118 = _____

11. 1445 − 978 = _____

12. 2136 − 1357 = _____

Multiply the following whole numbers:

13. 243 × 5 = _____

14. 125 × 6 = _____

15. 874 × 25 = _____

16. 719 × 74 = _____

17. 1863 × 134 = _____

18. 2085 × 236 = _____

Divide the following whole numbers:

19. 75 ÷ 5 = _____

20. 248 ÷ 4 = _____

21. 1450 ÷ 58 = _____

22. 867 ÷ 25 = _____

23. 936 ÷ 32 = _____

24. 425 ÷ 143 = _____

Find the lowest common denominator:

25. 1/2 and 1/4 = _____

26. 2/16 and 3/8 = _____

27. 1/16 and 3/4 = _____

28. 1/16 and 2/32 = _____

29. 1/2 and 1/16 = _____

30. 3/4 and 7/8 = _____

Reduce the following fractions:

31. 4/8 = _____

32. 8/16 = _____

33. 18/32 = _____

34. 16/64 = _____

35. 30/32 = _____

36. 42/64 = _____

Add the following fractions:

37. 1/4 + 3/4 = _____

38. 1/8 + 5/8 = _____

39. 3/16 + 1/4 = _____

40. 7/8 + 1/16 = _____

41. $2\ 3/8 + 4\ 1/8 =$ _____

42. $5\ 1/16 + 6\ 1/4 =$ _____

Subtract the following fractions:

43. $4/8 - 3/8 =$ _____

44. $7/32 - 5/32 =$ _____

45. $3/4 - 1/2 =$ _____

46. $7/8 - 1/4 =$ _____

47. $6\ 3/4 - 3\ 1/4 =$ _____

48. $7\ 1/2 - 4\ 7/8 =$ _____

Multiply the following fractions:

49. $2/4 \times 1/4 =$ _____

50. $1/4 \times 1/2 =$ _____

51. $3/8 \times 2/4 =$ _____

52. $1\ 1/4 \times 2\ 1/4 =$ _____

53. $6\ 1/2 \times 7/8 =$ _____

54. $1\ 3/8 \times 8 =$ _____

Divide the following fractions:

55. $3/4 \div 1/2 =$ _____

56. $2/8 \div 4/8 =$ _____

57. $1/8 \div 1/2 =$ _____

58. $1/4 \div 1/8 =$ _____

59. $2\ 1/4 \div 2\ 1/2 =$ _____

60. $3\ 1/2 \div 1\ 4/16 =$ _____

Add the following decimals:

61. $0.9 + 8.0 =$ _____

62. $7.8 + 7.9 =$ _____

63. $1.40 + 0.79 =$ _____

64. $2.46 + 3.59 =$ _____

65. $0.371 + 0.240 =$ _____

66. $0.337 + 0.262 =$ _____

Subtract the following decimals:

67. $0.8 - 0.2 =$ _____

68. $6.3 - 5.6 =$ _____

69. $0.27 - 0.13 =$ _____

70. $5.75 - 2.50 =$ _____

71. $0.860 - 0.034 =$ _____

72. $2.835 - 0.840 =$ _____

Multiply the following decimals:

73. $0.8 \times 2.0 =$ _____

74. $0.75 \times .50 =$ _____

75. $64.0 \times 6.2 =$ _____

76. $7.12 \times 5.25 =$ _____

77. $78.40 \times 12.00 =$ _____

78. $16.275 \times 1.232 =$ _____

Divide the following decimals:

79. $3.2 \div 0.4 =$ _____

80. $5.6 \div 0.7 =$ _____

81. $3.25 \div 1.25 =$ _____

82. $5.75 \div 0.5 =$ _____

83. $3.23 \div 1.70 =$ _____

84. $1.53 \div 17.00 =$ _____

Add the following dimensions:

85. 5 ft 4 in. + 2 ft 6 in. = _____

86. 8 ft 3 in. + 3 ft 5 in. = _____

87. 7 ft 10 in. + 4 ft 2 in. = _____

88. 6 ft 9 in. + 1 ft 6 in. = _____

89. 4 ft 7 in. + 3 ft 9 in. = _____

90. 2 ft 6 in. + 7 ft 7 in. = _____

Subtract the following dimensions:

91. 1 ft 8 in. − 6 in. = _____

92. 3 ft 9 in. − 2 ft 5 in. = _____

93. 4 ft 6 in. − 1 ft 6 in. = _____

94. 6 ft 7 in. − 3 ft 8 in. = _____

95. 8 ft 5 in. − 4 ft 7 in. = _____

96. 9 ft 2 in. − 5 ft 4 in. = _____

Multiply the following dimensions:

97. 6 ft × 2.5 ft = _____

98. 3 ft × 1.75 ft = _____

99. 2 ft 6 in. × 3 ft 4 in. = _____

100. 4 ft 3 in. × 2 ft 3 in. = _____

101. 5 ft 2 in. × 1 ft 1 in. = _____

102. 1 ft 3 in. × 2 ft 2 in. = _____

Divide the following dimensions:

103. 8 ft ÷ 2 = _____

104. 9 ft ÷ 3 = _____

105. 7 ft 6 in. ÷ 2 = _____

106. 6 ft 4 in. ÷ 2 = _____

107. 10 ft 6 in. ÷ 3 = _____

108. 2 ft 8 in. ÷ 4 = _____

Convert the following fractions to decimals:

109. 3/4 = _____

110. 1/2 = _____

111. 7/8 = _____

112. 1/16 = _____

113. 2/3 = _____

114. 2/5 = _____

Convert the following decimals to fractions:

115. Convert .50 to 4ths = _____

116. Convert .75 to 8ths = _____

117. Convert .25 to 16ths = _____

118. Convert .50 to 32nds = _____

119. Convert .25 to 8ths = _____

120. Convert .75 to 16ths = _____

Solve following the sequence of mathematical operations:

121. 2 × 6 ÷ 2 × 6 = _____

122. (2 × 6) ÷ (2 × 6) = _____

123. 6 × (5 + 4) = _____

124. (80 − 10) ÷ (2 + 8) = _____

125. 5×2^2 = _____

126. $6 \times 4 - 2^3$ = _____

CHAPTER 4

Reading Technical Drawings

Quiz

Name: _____

Date: _____

Class: _____

Instructor: _____

Grade: _____

Instructions: Carefully read Chapter 4 in the text and answer the following questions:

MULTIPLE-CHOICE

1. This item contains enough information to enable a welder to produce a weldment and may include a title page, pictorial drawing, assembly drawing, detailed drawing, and an exploded view.
   a. project routing information
   b. set of drawings
   c. bill of materials
   d. MSDS

2. Different types of lines are used to represent various parts of an object being illustrated, and these various line types are collectively known as
   a. orthographic projections
   b. specifications
   c. alphabet of lines
   d. pictorial drawings

3. These drawings are made as if you were looking through the sides of a glass box at an object and tracing its shape on the glass
   a. orthographic projections (mechanical drawings)
   b. pictorial drawings
   c. isometric drawings
   d. cavalier drawings

4. The views most commonly shown on mechanical drawings are
   a. rotated views
   b. dimension views
   c. back, left-side, and bottom views
   d. front, right-side, and top views

5. The numbers in this type of line give the size or length of an object.
   a. dimension lines
   b. extension lines
   c. cutting plane lines
   d. section lines

6. When a part must be drawn smaller or larger than it actually is, we use a
   a. section view
   b. cavalier drawing
   c. pictorial drawing
   d. drawing scale

7. This is the process of making a line on a drawing by making a series of quick, short strokes with the pencil or pen.
   a. dimensioning
   b. scaling
   c. sketching
   d. coloring

8. An easy scale to use is 1 in. equals 1 ft, so that if we draw a line that is 10 in. long, it represents an actual dimension of
   a. 10 in.
   b. 10 ft
   c. 1 ft
   d. 1 in.

9. These lines on a drawing show the edge of an object, the intersection of surfaces that form corners or edges, and the extent of a curved surface, such as the sides of a cylinder.
   a. object lines
   b. hidden lines
   c. centerlines
   d. extension lines

10. These make it easy to draw plans for a project that will be more precise and will allow you to move things around without having to erase.
    a. sketches
    b. drawings
    c. eraser shields
    d. computer drafting programs

SENTENCE COMPLETION

In the space provided write the answer that completes the statement.

1. Did you know that _____ are often called the universal language?

2. The two most common layouts for mechanical drawings are _____ _____ angle projection and _____ angle projection.

3. A group of drawings, known as a _____, should contain enough information to enable a welder to produce a weldment.

4. The _____ which appears in one corner of a mechanical drawing may contain the name of the part, company name, scale of the drawing, date of the drawing, who made the drawing, drawing number, number of drawings in the set, and tolerances.

5. The _____ detail the type and grade of material to be used, including base metal, consumables, such as filler metal, and hardware, such as nuts and bolts.

6. A _____ can also be included in the set of drawings and is a list of the various items that are needed to build the weldment.

7. Different types of lines are used to represent various parts of an object being illustrated and are known collectively as the _____.

8. _____ lines show the edge of an object, the intersection of surfaces that form corners or edges, and the extent of a curved surface, such as the sides of a cylinder.

9. _____ lines show the same features as object lines except that the corners, edges, and curved surfaces cannot be seen because they are hidden behind the surface of the object.

10. _____ lines show the center point of circles and arcs and round or symmetrical objects.

11. _____ lines are the lines that extend from an object and locate the points being dimensioned.

12. Numbers in the _____ line or next to it give the size or length of an object.

13. _____ lines represent an imaginary cut through the object and are used to expose the details of internal parts that would not be shown clearly with hidden lines.

14. _____ lines show the surface that has been imaginari-
ly cut away with a cutting plane line to show internal details.

15. There are two types of break lines: Long break lines and short break lines, and both
show that part of an object has been _____.

16. _____(the straight part) and _____
_____ (the pointed end) point to a part to identify it, show the location,
and/or are the basis of a welding symbol.

17. _____ lines show an alternate position of a moving
part or the extent of motion such as the on/off position of a light switch.

18. _____ drawings are made as if you were looking
through the sides of a glass box at the object and tracing its shape on the glass.

19. Pictorial drawings present the object in a more realistic or understandable form and
they usually appear as one of two types, _____ or
_____.

20. The front view is selected because when the object is viewed from this direction, its
overall _____ is best described.

21. _____ drawings are drawn at a 30° angle so it appears
you are looking at one corner.

22. On _____ drawings one surface, usually the front, is
drawn flat to the page.

23. The _____ view is made as if you were to saw away
part of the object to reveal internal details.

24. The _____ view is used to show detail within a part
that would be obscured by the part's surface.

25. _____ views are used to show small details of an area
on a part without having to draw the entire part larger.

26. _____ dimensions can be found on the front and
top views.

27. _____ dimensions can be found on the front and right-
side views.

28. _____ dimensions can be found on the top and right-
side views.

29. When we use a drawing _____, we are saying that the
part being drawn is smaller or larger than it really is.

30. Architectural scales are divided into fractions of inches; some of the common units are 1/8, 3/32, _____, 3/16, 3/8, and 3/4 inch.

31. Engineering scales are divided into _____ of inches; some of the common units are 10th, 20th, 30th, 40th, 50th, and 60th.

32. The standard views include the front, top, and _____ side views.

33. _____ is a quick and easy way of producing a drawing that can be used in the welding shop.

34. There are thin metal tools called _____ that are used in drafting to protect the neighboring line from erasure.

35. Making a sketch on _____ paper is a way of both making your drawing more accurate and speeding up the sketching process.

36. Computer drafting programs use _____ lines that are different from most drawing programs, which use raster art.

37. Because vector drawings are seen by the computer as lines, you can zoom in and out, measure, resize, reshape, or rotate the drawing, and the lines stay _____ _____.

38. Computers see raster images as a series of small squares called _____ _____.

Quiz

Name: _____

Date: _____

Class: _____

Instructor: _____

Grade: _____

Instructions: Carefully read Chapter 5 in the text and answer the following questions:

MULTIPLE-CHOICE

1. What term refers to the way pieces of metal are put together or aligned with each other?
   a. metal overlap
   b. specifications
   c. welding process
   d. weld joint design

2. The five basic joint designs are butt joints, lap joints, outside corner joints, edge joints, and _____.
   a. plate joints
   b. tee joints
   c. seam joints
   d. groove joints

3. What should be considered when choosing the weld joint design for a project?
   a. type and thickness of metal being welded
   b. the axis of the weld
   c. the surface finishing
   d. the root opening of the groove weld

4. What can be done before welding on thick plate or pipe to ensure that there is 100% penetration?
   a. use a plug weld
   b. control the distortion
   c. prepare the edge with a groove on one or both sides
   d. bring the edges together at an angle

5. What does a welding designer use on a drawing to indicate clearly to the welder the important detailed information regarding the weld?
   a. joint design
   b. code requirements
   c. welding symbols
   d. reinforcement

6. Five forces that act on a weld to cause stress are tensile, compression, bending, torsion, and _____.
   a. radius
   b. bevel
   c. faying
   d. shear

7. What can be added to a welding symbol for the placement of specific information?
   a. tail
   b. arrow
   c. angle
   d. triangle

8. The symbol for a fillet weld on a drawing is a
   a. rectangle
   b. right triangle
   c. circle
   d. square

9. What type of weld is made in the space cut into the joint between two pieces being joined?
   a. plug
   b. seam
   c. groove
   d. spot

10. What can be placed on the back side of a weld joint to prevent the molten metal from dripping through the open root?
    a. plug weld
    b. backing strip
    c. bevel
    d. reinforcement

SENTENCE COMPLETION

In the space provided write the answer that completes the statement.

1. The term _____ refers to the way pieces of metal are put together or aligned with each other.

2. The five basic joint designs are butt joints, lap joints, tee joints, _____
_____, and edge joints.

3. In a butt joint the edges of the metal _____ so that the thickness of the joint is approximately equal to the thickness of the metal.

4. In a lap joint the edges of the metal _____ so the thickness of the joint is approximately equal to the combined thickness of both pieces of metal.

5. Lap welds are usually joined by making a _____ weld along the edge of one plate joining it to the surface of the other.

6. In a _____ joint the edge of a piece of metal is placed on the surface of another piece of metal and the parts are usually placed at a 90° angle with each other.

7. In an _____ joint the edges of the metal are brought together at an angle, usually around 90° to each other.

8. In an edge joint the metal _____ are placed together so the edges are even.

9. Welding drawings and _____ usually tell you exactly which joint design will be used for all of the welds to be made.

10. When choosing which weld joint design to use, you must consider the type and thickness of the metal being welded, the welding position, welding process, finished weld properties, and any _____.

11. The forces acting on a weld can cause stresses in five ways: tensile, compression, bending, torsion, and _____.

12. The welding process to be used has a major effect on the selection of the _____
_____.

13. The area of the metal's surface that is _____ during the welding process is called the faying surface.

14. The faying surface can be shaped before welding to increase the weld's strength; this is called _____.

15. With the metal removed by _____ or beveling the metal's edge, it is easier for the molten weld metal to completely fuse through the joint.

16. A weld should be as strong as or stronger than the _____ being joined.

17. The edge of a thicker metal plate may be shaped with a bevel, V-groove, J-groove, or a _____.

18. When welding on thick plate or pipe, it is often impossible for the welder to get 100% _____ without some type of groove being used.

19. Back _____ is a process of cutting a groove in the back side of a joint that has been welded.

20. The most ideal welding position for most joints is the _____ _____ position because it allows for larger molten weld pools to be controlled.

21. Welds that are made in any position other than the flat position are referred to as being done _____.

22. In the _____ position for metal plate the welding is performed from the underside of the joint.

23. In the _____ position for pipe the pipe is vertical to the horizon, and the weld is made horizontally around the pipe.

24. Organizations such as the American Welding Society, American Society of Mechanical Engineers, and the American Bureau of Ships are a few of the agencies that issue _____.

25. Joint design can be a major way to control welding _____.

26. The information in the welding symbol can include details for the weld such as length, _____, height of reinforcement, groove type, groove dimensions, location, process, filler metal, strength, number of welds, weld shape, and surface finishing.

27. The _____ is added to the basic welding symbol when it is necessary to designate the welding specifications, procedures, or other supplementary information needed to make the weld.

28. A _____ weld is made by welding through a round hole in the top plate to fuse the bottom plate.

29. A _____ weld is approximately round and is created between the two overlapping surfaces being joined.

30. _____ welds are continuous along the overlapping surfaces and can be made by producing a series of overlapping spot welds or be one continuous resistance weld.

31. A _____ weld is made in the space cut into the joint between two pieces being joined.

32. A _____ is a piece of metal that is placed on the back side of a weld joint to prevent the molten metal from dripping through the open root.

33. Flange welds are used on thin metal as a way of stiffening the edge so there is less _____.

CHAPTER 6

Fabricating Techniques and Practices

Quiz

Name: _____

Date: _____

Class: _____

Instructor: _____

Grade: _____

Instructions: Carefully read Chapter 6 in the text and answer the following questions:

MULTIPLE-CHOICE

1. Why must extra safety precautions be taken during fabrication?
   a. fabrication is a more lengthy process
   b. several workers may be working on a structure at the same time
   c. most of the assembly is handmade
   d. preformed parts may be used

2. An advantage of using _____ parts for fabrication is that the cost per part is lower since they are mass produced rather than made one by one by hand
   a. automated
   b. custom
   c. preformed
   d. welded

3. An advantage of using _____ parts for fabrication is that sometimes people want to have something special or unique built or modified just for them.
   a. custom
   b. produced
   c. preformed
   d. low cost

4. It is important to know how to make the proper number, size, and location of tack welds because they
   a. are the final welds that will hold the assembly together
   b. are a major factor in the final cost of a weldment
   c. are the welds that hold the parts together so they can be finished welded
   d. cannot be corrected once they are placed

5. A good way to locate parts that are to be assembled is to
   a. line them up end to end
   b. face them all in one direction
   c. look at the direction that the metal is rolled
   d. line them up on an edge starting at a corner

6. One way to adjust parts so that they are within tolerance is to
   a. throw out the parts that do not fit and recut them
   b. change the tolerance
   c. loosen up the joint tolerance by slightly adjusting the alignment of each of the pieces
   d. parts cannot be adjusted to meet the tolerance

7. One way to control metal distortion during welding is to
   a. increase the heat input
   b. bend the metal away from the heat
   c. make long welds
   d. make small uniform welds

8. Laying out parts together in a manner that will minimize the waste created is called
   a. shaping
   b. curving
   c. fabricating
   d. nesting

9. The most common shapes of metal used in fabrication are plate, sheet, tubing, angles, and _____.
   a. pipe
   b. circles
   c. stock
   d. squares

10. The first step in the assembly process is to
    a. adjust the tolerance
    b. select the largest or most central part to be the base for your assembly
    c. mark layout lines and other markings
    d. tack weld the parts

SENTENCE COMPLETION

In the space provided write the answer that completes the statement.

1. The first step in almost every welding operation is the _____ _____ of the parts to be joined by welding.

2. The difference between a weldment and a fabrication is that a _____ _____ is an assembly whose parts are all welded together, but a _____ _____ is an assembly whose parts may be joined by a combination of methods including welds, bolts, screws, adhesive, and so forth.

3. Welded fabrications can be made from precut and preformed parts, or they can be made from parts _____.

4. When making an assembly with precut and formed parts, little or no on-the-job _____ may be required.

5. The opposite end of the spectrum from preformed parts is _____ _____ fabrication, in which all or most of the assembly is handmade.

6. Often, even if there are going to be thousands or even tens of thousands of a weldment produced, the first one, the _____, must be made by hand to be sure that everything works as it was planned.

7. _____ welds are the welds, usually small in size, that are made during the assembly to hold all of the parts of a weldment together so they can be finished welded.

8. A large number of very small tack welds should be used on _____ _____ metal sections, while a few large tack welds may be used for _____ _____ metal parts.

9. Short joints take fewer welds, but some long, straight joints may have very few tack welds compared to a shorter joint that is very _____.

10. The more exacting the _____ for the finished weldment, the more tack welds are required.

11. Tack welds must be located well within the joint so that they can be completely _____ into the finished weld.

12. A good tack weld is one that does its job by holding parts in place yet is _____ _____ in the finished weld.

13. Locating parts is easier when the parts being assembled are lined up on an _____ _____ starting at a corner.

14. A part's _____ is the amount that a part can be bigger or smaller than it should be and still be acceptable.

15. As the number of parts that make up a weldment increases, the problem of compounding the errors _____.

16. Whenever possible, try to get the parts to fit without having to recut or _____ _____ them; but remember, the finished weldment must be within tolerance.

17. If the finished weldment is not within _____, it may be unusable.

18. Larger welds result in more _____ being added and more heat input to the base metal.

19. All metals distort by expansion when _____ and distort by contraction when _____.

20. Parts return to their original _____ when cooled if the heating is uniform and their shapes symmetrical.

21. The two factors that affect the degree to which a metal will distort and possibly remain distorted are its rate of _____ and its rate of _____.

22. Basically, the higher the coefficiency of _____, the greater the metal distorts.

23. All metals expand when _____ and contract when _____.

24. The more uniform a part is heated, the more uniform the expansion and contraction, which can result in the metal being less _____ following the heat-cooling cycle.

25. As metal is heated, it initially bends away from the _____, but when it is allowed to cool, it bends even farther back toward the heated spot.

26. As weld metal cools, it _____ or pulls together, causing distortion.

27. Welds that are more uniform in shape or symmetrical like U-groove welds cause the metal to expand uniformly and contract uniformly, resulting in less _____.

28. _____ and welding metal markers are made to withstand most welding and cutting temperatures without vanishing like a line from a felt-tip marker will.

29. The end of the soapstone should be _____ to increase accuracy.

30. A _____ line will make a long, straight line on metal and is best used on large jobs with long, straight lines.

31. Either a scribe or a punch can be used to lay out an accurate line, but the _____ _____ line can be easier to see when cutting.

32. To avoid making a cutting _____, lines should be identified as to whether they are being used for cutting, locating bends, drill centers, or assembly locations.

33. One advantage for most welding assemblies is that many errors in cutting can be repaired by _____.

34. Parts with square and straight cuts are easier to lay out than are parts with _____ _____.

35. The placement of parts together in a manner that will minimize the waste created is called _____.

36. All cutting processes except shearing remove some metal, leaving a small gap or space called the _____.

37. Of the cutting processes used in most shops, the metal saw produces one of the smallest kerfs while the handheld oxyfuel _____ can produce one of the widest.

38. An advantage to tracing a part or template is that it is fast and easy to mark out a large number of parts that are exactly the _____.

39. To reduce the error caused by tracing, be sure the line you draw is made _____ _____ to the part's edge.

40. The most common metal used is _____, and the most common shapes used are plate, sheet, pipe, tubing, and angles.

41. To start the assembly, make a selection of the largest or most _____ _____ part to be the base for your assembly; all other parts will be aligned to this one part.

42. _____ are one of the most commonly used clamps.

43. _____ are devices that are made to aid in assemblies and fabrication of weldments and may align, position, or support parts.

44. A simple way to correcting slight alignment problems is to make a small _____ _____ in the joint and then use a hammer and possibly an anvil to pound the part into place.

CHAPTER 7
Welding Shop Practices

Quiz

Name: _____
Date: _____
Class: _____
Instructor: _____
Grade: _____

Instructions: Carefully read Chapter 7 in the text and answer the following questions:

MULTIPLE-CHOICE

1. As an employee, good companies will be judging you not only by your welding skills but also by your
   a. efficiency and productivity
   b. age
   c. family background
   d. the people you know

2. A good way to more easily get any assistance you need on the job is to
   a. always ask the boss
   b. ask many questions
   c. not bother others with questions
   d. develop a friendly working relationship with coworkers

3. What is an acceptable way to improve a company's profit margin and ability to compete when bidding on jobs?
   a. conserve materials and supplies
   b. reduce the number of welders
   c. bid high
   d. use lower quality materials

4. Reusing leftover material from a previous job and nesting parts during layout are ways to
   a. introduce errors into the fabrication
   b. speed up the layout process
   c. increase costs
   d. save on the cost of materials

5. One acceptable way to save on the cost of shielding gas is to
   a. use a higher gas pressure
   b. set the gas pressure as low as possible within the manufacturer's recommended pressures
   c. lower the voltage
   d. set the flow rate below the manufacturer's recommended flow rate

6. Turning off welding machines when they are not being used is a way to
   a. reduce the time required to weld
   b. waste electricity
   c. increase the noise level
   d. save energy

7. A way to help the environment and be a source of revenue for the welding shop is to
   a. clean up your welding area
   b. pick up parts from a welding supply
   c. sell scrap material to a recycler
   d. take home leftover materials

8. Damp electrodes can cause
   a. a lower welding current
   b. the wrong polarity
   c. excessive spatter
   d. too long an arc length

9. The best way to ensure the safe operation of equipment in the shop is to
   a. have someone remind you of the rules
   b. read and follow all manufacturers' operating and safety instructions
   c. use your own judgment as to what is safe
   d. work slowly

10. When would it be appropriate to use hand signals to communicate on the job?
    a. while welding
    b. when communicating with your boss
    c. when communicating with crane operators
    d. it is never appropriate

SENTENCE COMPLETION

In the space provided write the answer that completes the statement.

1. How well you work with _____ is as important as how well you can weld.

2. Your _____ is your ability to get jobs done and make the company profitable.

3. Developing a friendly working relationship with _____ will allow you to more easily ask for help.

4. Often a job will go _____ if you work with one or more people.

5. At the beginning of the day, you may want to spend a few minutes talking with your boss and _____ about how to best approach a project so that it will move through the shop most efficiently.

6. Labor costs, the money your employer spends on your salary, or money you earn as a result of your work in your own business is a major cost factor in the _____ _____.

7. Sometimes the total time spent on a project can be reduced by building a _____ _____ that can help align parts and tack them in place for welding.

8. When you are _____, the job you were scheduled to do for the day may not get done.

9. Being _____ is one of the many factors that employers consider when deciding about pay raises, job advancements, and in some cases, continued employment.

10. It is important to have a friendly and professional work environment, but _____ _____ and _____ have no place on the job.

11. One way to reduce material cost is to look for ways to reuse _____ _____ from a previous job.

12. Another important way to save on material costs is to make sure you are being as conservative as possible with materials when you are _____ the parts on new stock.

13. Even with a short stub, approximately _____% of the total electrode weight purchased is lost due to losses of spatter, slag, fumes, and the stub.

14. Sometimes partially used electrodes may cause a small _____ _____ in the weld as they are restarted.

15. Positioning the work so that as much as possible of the welding can be done in the _____ position makes it easier on the welder, more comfortable for the welder, and increases the welding productivity.

16. Never continue to weld when you observe excessive _____ being produced.

17. Shielding gases are used to protect the molten weld metal from _____ _____.

18. When possible, set the shielding gas flow rate on GMA, GTA, and FCA welding processes at the _____ end of the flow rate range to reduce the shielding gas cost.

19. Both too low a gas pressure and too high a gas pressure result in _____ _____.

20. All welding and cutting processes except for the few oxyfuel processes use large quantities of _____.

21. The best way of saving energy when using any welding equipment is to _____ _____ the machines when they are not being used.

22. _____ weld shop materials is important because it both helps the environment and can be a source of revenue for the shop as the scrap is sold to a recycler.

23. It is important to read and follow all _____ operating and safety instructions for each piece of equipment.

24. _____ can provide the necessary instruction for the crane operator to safely and accurately place a component.

25. Some of the reasons a shop might _____ are because the project requires equipment your shop does not have or another shop can produce it more cost effectively.

CHAPTER 8

Shielded Metal Arc Equipment, Setup, and Operation

Quiz

Name: _____

Date: _____

Class: _____

Instructor: _____

Grade: _____

Instructions: Carefully read Chapter 8 in the text and answer the following questions:

MULTIPLE-CHOICE

1. In shielded metal arc welding what is used to carry the electrical current?
   a. gaseous cloud
   b. base metal
   c. flux covered metal electrode
   d. amperage

2. What three units are used to describe any electrical current?
   a. electrons, arc, temperature
   b. voltage, amperage, wattage
   c. temperature, heat, voltage
   d. alternating, anode, cathode

3. What are the three types of welding current used in SMA welding?
   a. AC, DC, alternating
   b. positive, negative, reverse
   c. voltage, wattage, amperage
   d. DCEN, DCEP, AC

4. What kind of voltage is much like the higher surge of pressure you might observe when a water hose nozzle is first opened?
   a. operating
   b. closed circuit
   c. welding
   d. open circuit

5. What term is used to describe when the arc moves between the electrode and the work during welding due to unevenly spaced flux lines?
   a. magnetic fields
   b. operating voltage
   c. arc blow
   d. constant current

6. What type of transformer takes a high-voltage, low-amperage current and changes it into a low-voltage, high-amperage current?
   a. step-up
   b. multiple coil
   c. inverter
   d. step-down

7. What type of welders produce welding electricity from a mechanical power source?
   a. inverter-type
   b. generators and alternators
   c. movable core
   d. tap-type

8. A 70% duty cycle means that out of every 10 minutes, the machine can be used for how many minutes at the maximum rated current?
   a. 3
   b. 10
   c. 7
   d. 70

9. When setting up a welding workstation, the area must not contain
   a. combustible materials
   b. welding cables
   c. ladders
   d. dust

10. Welding cables in the workstation should be
    a. long enough to reach the workstation
    b. coiled
    c. tied to scaffolding or ladders
    d. wrapped around your arm or other part of your body

SENTENCE COMPLETION

In the space provided write the answer that completes the statement.

1. Shielded metal arc welding (SMAW) is a welding process that uses a flux covered metal electrode to carry an _____ current.

2. The current forms an _____ across the gap between the end of the electrode and the work.

3. The metal from the electrode and the molten _____ are mixed together to form the weld.

4. The high temperature at the electrode end causes the flux covering around the electrode to burn or vaporize into a _____.

5. Some of the electrode flux forms a molten protective _____ on top of the molten weld pool.

6. _____ is the most widely used welding process for metal fabrication because of its low cost, flexibility, portability, and versatility.

7. *Welding current* is the term used to describe the _____ that jumps across the arc gap between the end of the electrode and the metal being welded.

8. An electric current is the flow of _____.

9. The greater the electrical resistance, the greater the _____ and _____ the arc will produce.

10. Voltage controls the maximum gap that the electrons can jump to form the _____ _____.

11. Amperage controls the _____ of the arc.

12. _____ is a measurement of the amount of electrical energy or power in the arc.

13. The shorter the arc, the lower the arc voltage and the lower the _____ _____ produced, and as the arc lengthens, the resistance increases, thus causing a rise in the arc voltage and temperature.

14. The amount of heat produced by the arc is determined by the _____ _____.

15. In total, about _____% of all heat produced by an arc is missing from the weld.

16. In direct current electrode negative, the electrode is negative, and the work is _____.

17. The electrons are leaving the _____ and traveling across the arc to the surface of the metal being welded.

18. In direct current electrode positive, the electrode is positive, and the work is _____ _____.

19. The electrons are leaving the surface of the metal being welded and traveling across the arc to the _____.

20. In _____ current, the electrons change direction every 1/120 of a second so that the electrode and work alternate from anode to cathode.

21. _____ voltage is the voltage at the electrode before striking an arc (with no current being drawn).

22. The maximum safe open circuit voltage for welders is _____ V.

23. Operating, welding, or _____ voltage, is the voltage at the arc during welding.

24. The best way to eliminate arc blow is to use _____ current.

25. A _____ transformer takes a high-voltage, low-amperage current and changes it into a low-voltage, high-amperage current.

26. The _____ machine, or tap-type machine, allows the selection of different current settings by tapping into the secondary coil at a different turn value.

27. _____ or movable core machines are adjusted by turning a handwheel that moves the internal parts closer together or farther apart.

28. _____ welding machines are much smaller than other types of machines of the same amperage range.

29. The use of electronics in the _____-type welder allows it to produce any desired type of welding power.

30. An _____ can produce AC only.

31. A _____ can produce AC or DC.

32. Alternating welding current can be converted to direct current by using a series of _____.

33. The _____ cycle is the percentage of time a welding machine can be used continuously.

34. A 60% duty cycle means that out of every 10 minutes, the machine can be used for _____ minutes at the maximum rated current.

35. Cables used for welding must be flexible, well insulated, and the correct _____ _____ for the job.

36. A properly sized electrode holder can _____ if the jaws are dirty or too loose, or if the cable is loose.

37. The work clamp must be the correct size for the _____ being used, and it must clamp tightly to the material.

38. Arc welding machines should be located near the welding site but far enough away so that they are not covered with _____.

39. The welding machine and its main power switch should be _____ _____ while a person is installing or working on the cables.

CHAPTER 9

Shielded Metal Arc Welding Plate

Quiz

Name: _____

Date: _____

Class: _____

Instructor: _____

Grade: _____

Instructions: Carefully read Chapter 9 in the text and answer the following questions:

MULTIPLE-CHOICE

1. Which of the following is personal protective equipment (PPE) required for shielded metal arc welding safety?
   a. chipping hammer
   b. electrode holder
   c. welding helmet
   d. grinder

2. Which electrode can be used on metal that has a little rust, oil, or dirt without seriously affecting the weld's strength?
   a. 6011
   b. 6013
   c. 7018
   d. 7020

3. Which electrode has a rutile-based flux, giving a smooth, easy arc with a thick slag left behind on the weld bead?
   a. 6011
   b. 6013
   c. 7018
   d. 7020

4. An electrode angle that pushes molten metal and slag ahead of the weld is the
   a. trailing electrode angle
   b. perpendicular angle
   c. leading electrode angle
   d. backhand angle

5. Welding in a comfortable position will
   a. have no effect on weld quality
   b. cause welder fatigue
   c. feel awkward
   d. improve weld quality

6. The distance that the arc must jump from the end of the electrode to the plate or weld pool surface is the
   a. arc length
   b. stand-off distance
   c. electrode gap
   d. electrode space

7. Setting too low a shielded metal arc welding current for an electrode type and size can result in
   a. excessive spatter
   b. poor fusion
   c. wider weld beads
   d. weld burnthrough

8. The purpose of a tack weld is to
   a. hold small parts in place
   b. help you set a proper welding current
   c. temporarily hold parts in place until they can be completely welded
   d. fix a weld bead burnthrough

9. Ideally, the welder should strike the arc
   a. wherever it is convenient
   b. within ½ in. of the weld joint
   c. no further than ½ in. from the weld joint
   d. in the weld joint just ahead of where the weld is going to be made

10. Welds that are made in a straight line with little or no side-to-side electrode movement are
    a. tack welds
    b. stringer beads
    c. surface welds
    d. butt welds

SENTENCE COMPLETION

In the space provided write the answer that completes the statement.

1. Arc welding electrodes used for practice welds are grouped into three filler metal classes (F number) according to their major welding characteristics. The groups are _____.

2. F3 E6010 and E6011 electrodes—Both of these electrodes have _____ _____ fluxes.

3. E6010 and E6011 are the most utilitarian welding electrodes for welding fabrication. They can be used on metal that has a little _____ without seriously affecting the weld's strength.

4. F2 E6012 and E6013 electrodes—These electrodes have _____ fluxes, giving a smooth, easy arc with a thick slag left on the weld bead.

5. F4 E7016 and E7018 electrodes—Both of these electrodes have a _____ _____ flux.

6. The electrodes with the cellulose-based fluxes do not have heavy _____ _____ that may interfere with the welder's view of the weld.

7. Electrodes with the _____ fluxes (giving an easy arc with low spatter) are easier to control and are used for fillet, stringer beads, and butt joints.

8. Welding with the current set too _____ results in poor fusion and poor arc stability.

9. The _____-in. (_____-mm) electrode is the most commonly used size for metal fabrication.

10. The _____ rate, or the rate that weld metal is added to the weld, is slower when small diameter electrodes are used.

11. Large diameter electrodes may _____ the metal if they are used with thin or small pieces of metal.

12. The _____ length is the distance the arc must jump from the end of the electrode to the plate or weld pool surface.

13. To maintain a constant arc length, the electrode must be _____ _____ continuously.

14. Long arc lengths produce more _____ because the metal being transferred may drop outside of the molten weld pool.

15. An arc will jump to the closest _____ conductor.

16. Because _____ arcs produce less heat and penetration, they are best suited for use on thin metal or thin-to-thick metal joints.

17. Most welding jobs require an arc length of _____ in. (_____ mm) to 3/8 in. (10 mm), but this distance varies.

18. The electrode angle is measured from the electrode to the _____.

19. A _____ electrode angle pushes molten metal and slag ahead of the weld.

20. A _____ electrode angle pushes the molten metal away from the leading edge of the molten weld pool toward the back where it solidifies.

21. The movement or weaving of the welding electrode, called _____ _____, can control the following characteristics of the weld bead: penetration, buildup, width, porosity, undercut, overlap, and slag inclusions.

22. Welders must have enough freedom of movement so that they do not need to change _____ during a weld.

23. When welding, even if a welder is seated, touching a _____ object will make the welder more stable and will make welding more relaxing.

24. All welds start with an _____. It is the process of establishing a stable arc between the end of the electrode and the work.

25. One important thing to remember is that on most code welding jobs, an arc strike _____ of the weld zone may be considered a defect.

26. If the electrode sticks to the plate, quickly squeeze the electrode holder lever to _____ the electrode.

27. Tack welds are a _____ method of holding parts in place until they can be completely welded.

28. A straight weld bead on the surface of a plate with little or no side-to-side electrode movement is known as a _____.

29. The root opening for most butt welds varies from 0 to _____ in. (3 mm).

30. An outside corner joint is made by placing the plates at an angle to each other, with the edges forming a _____.

31. A lap joint is made by _____ the edges of the two plates.

32. As the fillet weld is made on the lap joint, the buildup should equal the _____ _____ of the plate.

33. The tee joint is made by tack welding one piece of metal on another piece of metal at a _____ angle.

Shielded Metal Arc Welding Pipe

Quiz

Name: _____

Date: _____

Class: _____

Instructor: _____

Grade: _____

Instructions: Carefully read Chapter 10 in the text and answer the following questions:

MULTIPLE-CHOICE

1. What material is used to both carry fluids and for structural applications?
   a. pipe
   b. metal plate
   c. angle iron
   d. I-beam

2. Pipe that is 12 in. or smaller in diameter is measured by its
   a. outside diameter
   b. radius
   c. perimeter
   d. inside diameter

3. To ensure high joint strength, the ends of larger diameter pipe are usually
   a. graded
   b. beveled
   c. flat
   d. threaded

4. A characteristic of welded fittings for structural items is that
   a. they are more expensive
   b. any angle fitting can be fabricated
   c. forged fittings are limited
   d. only joints in small diameter pipe can be fabricated

5. The difference between a welded pipe tee joint for fluid and one for a structural item is that the structural fitting
   a. does not have to be welded as well
   b. must be welded better
   c. is lighter
   d. does not need to have the center cut out

6. What is the first weld made in a pipe joint?
   a. root
   b. filler
   c. hot pass
   d. cap

7. A primary purpose for the root face is to
   a. control the bead face
   b. help align the pipe joint
   c. control penetration
   d. make it easier to run the hot pass

8. What is the most important part of a root weld on a pipe?
   a. the outside surface
   b. the weld face
   c. the weld height
   d. the inside surface

9. To ensure good fusion, pipe weld bead starts and stops should be
   a. as large as possible
   b. as small as possible
   c. staggered
   d. at the same spot

10. What pipe position is commonly used in fabrication shops where the structure can be repositioned during welding?
    a. 1G
    b. 2G
    c. 3G
    d. 4G

SENTENCE COMPLETION

In the space provided write the answer that completes the statement.

1. Most piping systems use _____ fittings while structural items use shop-fabricated fittings.

2. Another difference is that in most cases there is no reason to cut out the center of a structural fitting because nothing will be _____ through the pipes.

3. Pipe is primarily used to carry _____, but is also used for structural applications.

4. Tubing can be used for both but is primarily used for _____ applications.

5. The specifications for pipe sizes are given as the _____ diameter (ID) for pipe 12 in. (305 mm) in diameter or smaller and as the _____ _____ diameter (OD) for pipe larger than 12 in. (305 mm) in diameter.

6. The wall thickness for pipe is determined by its _____.

7. Tubing sizes are always given as the _____ diameter.

8. Pipe and tubing are both available as welded (seamed) or _____ _____ (seamless).

9. The end of a pipe may be cut square, or it may be _____.

10. Larger diameter pipe and that being used in a piping system are usually _____ _____ for maximum penetration and high joint strength.

11. It is important that the bevel be at the correct angle, about _____ _____° depending on specifications, and that the end meet squarely with the mating pipe.

12. Root suck back is caused by the surface tension of the molten metal trying to pull itself into a ball, forming a _____ root surface.

13. Fitting pipe together and holding it in place for welding become more difficult as the diameter of the pipe gets _____.

14. Pipe used for piping systems is often welded using either _____ _____ electrodes for the complete weld.

15. _____ pipe welds are made up of several separate weld passes.

16. A root weld is the _____ weld in a joint.

17. The most important part of a root weld is the internal root face or, in the case of pipe, the _____ surface.

18. The hot pass is used to quickly burn out small amounts of _____ _____ trapped along the edge of the root pass.

19. The filler pass(es) may be either a series of stringer beads, Figure 10-17, or a _____ bead.

20. When the bead has gone completely around the pipe, it should continue past the _____ point so that good fusion is ensured.

21. The final covering on a grooved weld is referred to as the _____ _____ pass or cap.

22. Cover passes that are excessively _____ will reduce the pipe's strength, not increase it.

23. The _____ pipe position is commonly used in fabrication shops where structures or small systems can be positioned for the convenience of the welder.

24. A weld being produced on a horizontal pipe is being made in flat, vertical up or vertical down, and _____ positions.

25. When making the root pass on a pipe joint, the welding direction is usually made based on the _____.

26. _____ horizontal pipe is the easiest and fastest way of making horizontal pipe welds.

27. A bevel extends all the way across the part's surface; a _____ removes only a corner of the surface.

CHAPTER 11

Gas Metal Arc Welding Equipment and Materials

Quiz

Name: _____

Date: _____

Class: _____

Instructor: _____

Grade: _____

Instructions: Carefully read Chapter 11 in the text and answer the following questions:

MULTIPLE-CHOICE

1. The electrode used in GMA welding is fed
   a. manually
   b. by gravity
   c. automatically
   d. semi automatically

2. Why does GMA welding use a shielding gas?
   a. to prevent spatter
   b. to cool the weld
   c. to make it easier to see the weld
   d. to protect the weld from the atmosphere

3. A wire feeder, power source, work clamp, and shielding gas source are all
   a. commonly replaceable parts
   b. used to provide the current for the arc
   c. part of a GMA welding station
   d. painted black so they do not reflect arc light

4. The best shielding gas for making welds on thick sections of aluminum would be
   a. CO_2
   b. CO_2 plus argon
   c. argon plus 10% oxygen
   d. helium

5. The best gas flow rate should be
   a. 35 CFH
   b. 35 CFM
   c. as high as possible
   d. selected by using welding guides

6. In the short-circuit method of metal transfer the electrode
   a. forms a ball of molten metal that drops into the weld
   b. comes in contact with the base metal surface
   c. is vaporized by the arc
   d. forms a fine stream of molten droplets

7. Globular transfer is generally used
   a. on thin materials at a very low current range
   b. only in the flat position
   c. on large diameter pipe
   d. on thick material with a high current range

8. The purpose of the cast in the electrode wire is to
   a. make it easier to thread the wire
   b. ensure a positive contact with the energized contact tube
   c. keep the filler metal centered in the weld pool
   d. prevent long pieces of scrap wire from becoming a trip hazard

9. The American Welding Society's method of identifying GMA welding electrodes uses a series of letters and numbers that indicate the
   a. chemical and physical properties of the weld produced by the filler metal
   b. specific composition of the wire
   c. cost of the wire
   d. length of the wire

10. What could make welding out of position impossible?
    a. using argon 2% oxygen gas
    b. using 100% CO_2
    c. an improperly set up system
    d. using 0.45 diameter wire

SENTENCE COMPLETION

In the space provided write the answer that completes the statement.

1. GMA welding uses a solid welding wire that is fed _____ at a constant speed as an electrode.

2. During the welding process, a shielding gas protects the weld from the atmosphere and prevents _____ of the base metal.

3. This process is often referred to by its original name, MIG, which stands for _____ _____.

4. The same size filler metal and type of shielding gas can be used to make welds on thin or thick metal by simply adjusting the _____.

5. GMA welding can easily be used in _____ positions because the weld pool is small and the metal is molten for a very short time.

6. The main piece of equipment is the power source, which is often called the _____ _____.

7. Because of the long periods of continuous use, GMA welding machines have a _____ _____% duty cycle.

8. The terms *voltage, volts,* and *potential* can all be interchanged when referring to electrical _____.

9. The terms a*mperage, amps,* and *current* can all be interchanged when referring to electrical _____.

10. GMA welding power supplies produce a _____ arc voltage.

11. The purpose of the _____ is to provide a steady, uniform, and reliable supply of wire to the weld.

12. _____ results when the wire momentarily stops feeding, and the arc backs up to fuse the welding contact tip, resulting in damage to the equipment.

13. U-grooved rollers should be used when feeding _____ wires.

14. V-grooved rollers are best suited for _____ wires, such as mild steel and stainless steel.

15. The groove in the feed roller must be properly sized to fit the _____ _____ diameter being fed.

16. In the push-type system, the small diameter electrode wire must have enough _____ _____ to be pushed through the conduit without kinking.

17. In pull-type systems, a small motor is located in the gun to _____ _____ the wire through the conduit.

18. The push-pull–type wire-feed system has two _____, one located at each end.

19. A _____ gun is a compact, self-contained wire-feed unit consisting of a small drive motor and a wire supply.

20. The feed rollers clamp onto the filler wire, and as the rollers spin, they push or pull the wire from the reel to the _____.

21. For the feed rollers to properly grip and feed the filler metal, they must have proper _____ applied.

22. To prevent coasting, the tension on the _____ must be properly set.

23. The _____ guide is a small pointed tube that guides the wire from the feed rollers into the beginning of the wire liner.

24. The conduit liner is a flexible hollow tube through which the wire is fed from the wire-feed unit to the _____.

25. The main part of the GMA gun is called the _____.

26. The gun trigger is a _____ switch (off/on switch) and is attached to the gun body.

27. The insulated conductor tube, sometimes called the _____, is attached to the body of the gun.

28. The most commonly used conductor tubes have a _____° angle.

29. The gas diffuser allows the _____ to be dispersed or diffused around the contact tube, enveloping the entire weld area for greater protection.

30. The contact tube is a short replaceable copper conductor that transfers the electricity from the gun to the _____.

31. To keep the electrically hot components from accidentally arcing, an insulating device called the gun nozzle _____ is used.

32. The _____ is the hollow metal tube located at the end of the gun assembly through which the welding wire passes and the shielding gas flows.

33. To be sure that you have a good work connection, remove any paint, dirt, rust, oil, or other surface contamination at the point that the _____ is connected to the weldment.

34. The shielding gas _____ measures the flow rate in cubic feet per hour (CFH) or in metric measure as liters per minute (L/min).

35. The GMA shielding gas can be provided from a _____ or from a central gas piped manifold system.

36. Inert gases do not react with any other substance and are _____ _____ in molten metal.

37. Because argon is _____ than air, it effectively shields welds by pushing the lighter air away.

38. Helium is _____ than air, thus its flow rates must be about twice as high as argon's for acceptable stiffness in the gas stream to be able to push air away from the weld.

39. One hundred percent _____ is widely used as a shielding gas for GMA welding of steels.

40. The chief drawback in the use of carbon dioxide is the less steady-arc characteristics and a considerable increase in weld _____.

41. One hundred percent _____ can be used to weld copper and copper alloys.

42. High gas flow rates both waste shielding gases and may lead to _____ _____.

43. The larger the nozzle size, the higher the permissible flow rate without causing _____.

44. _____ provides the highest metal transfer rate and that means higher productivity.

45. The mode of metal transfer is the mechanism by which the molten filler metal is transferred across the arc to the _____.

46. The modes of metal transfer are short-circuit transfer (GMAW-S), axial spray transfer, globular transfer, and _____.

47. To change from one metal transfer mode to another, all you have to do is make the necessary changes in the voltage and _____ settings.

48. In short-circuit metal transfer, the electrode actually momentarily comes in contact with the _____, and the arc is momentarily shorted out.

49. The _____ mode of transfer is the most common process used with GMAW.

50. Globular transfer is generally used on _____ materials and at a very low current range.

51. The GMA axial spray metal transfer mode uses the highest _____ _____ and amperage settings compared to other processes.

52. The axial spray transfer process is very hot and virtually free of any _____ _____.

53. A disadvantage of axial spray transfer is that it produces a very fluid weld pool that may be difficult to _____ in out-of-position welds.

54. A _____ filter lens is required when using the spray transfer mode.

55. The high current of the spray mode produces good penetration and fusion, and the low current of the _____ transfer allows the weld pool to cool and contract slightly, so it is easier to control.

56. The electrode used with GMAW is a very long coil of _____ wire.

57. The _____ is the diameter of the circle formed by an unrestricted coil of wire.

58. The _____ is the twist in the wire.

59. The slight bend of the cast in the electrode wire ensures a positive contact with the energized _____.

60. Oxidation or even rusting may occur if the electrode is stored in a _____ _____ location.

61. For example, ER70 indicates that the weld metal, as deposited, will have a minimum tensile strength of _____ pounds per square inch.

62. With ER70S-2, the 2 shows the electrode is a _____ mild steel filler wire and can be used on metal that has a light cover of rust or oxide.

63. With ER70S-3, the 3 indicates this electrode does not have the _____ _____ needed for welding over a metal surface that has rust or mill scale on it.

64. With ER70S-4, the 4 shows that the wire electrode contains a _____ _____ level of a deoxidizer like silicone than ER70S-3.

65. With ER70S-5, the 5 shows that this electrode is used in the _____ _____ position only.

66. With the ER70S-6, the 6 shows this is an electrode with the highest levels of manganese and silicone for strength and _____.

67. With the ER70S-7, the 7 shows this is a high manganese carbon steel electrode with a balanced level of _____.

68. With the ER70S-G, the G indicates that this is a _____ electrode.

69. If the shielding gas supply is a cylinder, it must be chained securely in place before the _____ is removed.

70. Check the feed roller size to ensure that it matches the _____ size.

71. The conduit or an extension should be aligned with the groove in the roller and set as close to the roller as possible without _____.

72. _____ of the electrode wire results when the feed roller pushes the wire into a tangled ball because the wire would not go through the outfeed side conduit, with the wire resembling a bird's nest.

73. Loose fittings can leak; loose connections can cause added _____ _____, reducing the welding efficiency.

74. Pressing the gun switch to start the wire feeder is called _____ _____ the gun.

Gas Metal Arc Welding

Quiz

Name: _____
Date: _____
Class: _____
Instructor: _____
Grade: _____

Instructions: Carefully read Chapter 12 in the text and answer the following questions:

MULTIPLE-CHOICE

1. What affect does changing the wire-feed speed have on GMA welding?
   a. it changes the voltage
   b. it changes the amperage
   c. it affects the slope
   d. it does not affect the weld

2. How can wire-feed speed be measured?
   a. by timing and measuring the length of wire
   b. by counting the number of turns the wire spool makes in one minute
   c. by multiplying the diameters of the drive roller by 3.14 (pie) and timing its speed
   d. by multiplying the spool diameter by 3.14 (pie) and its RPM

3. At high wire-feed speeds, a 6-second test may be preferable to a 1-minute test because
   a. it is more accurate
   b. it saves times
   c. less wire will be fed out
   d. only one-minute tests are used for setting the wire-feed speed

4. If the wire-feed speed increases, the welding
   a. current will decrease
   b. voltage will increase
   c. current will increase
   d. voltage will decrease

5. An advantage of using the forehand welding technique is
   a. the depth of the weld penetration
   b. the fast rate of travel
   c. no spatter
   d. you can easily see the joint where the bead will be deposited

6. What welding gun angle does most machine and robotic weldings use?
   a. backhand angle
   b. perpendicular angle
   c. forehand angle
   d. pushing angle

7. What is the major effect that changing the electrode extension has on the weld?
   a. it reduces the wire-feed speed
   b. it affects the quantity of welding fumes produced
   c. it improves the weld bead visibility
   d. it changes the weld penetration

8. What effect does changing the gun angle have on the weld?
   a. weld depth and joint penetration
   b. increases the weld spatter
   c. decreases the weld spatter
   d. has little or no effect on the weld

9. To make a quality weld you must
   a. make changes in the welding setup as the weld progresses
   b. keep the speed, weave, gun angle, and electrode extension constant
   c. make the weld in a fixed position
   d. wear proper clothing so you do not jump if a spark lands on your hand

10. A crack can be stopped from going all the way along a joint by
    a. making bigger welds
    b. preheating the weld metal
    c. making intermittent welds
    d. using a stronger filler weld metal

SENTENCE COMPLETION

In the space provided write the answer that completes the statement.

1. Making a good GMA weld requires that you have both the welding skills and knowledge to properly _____ the welding machine.

2. The knowledge to set up the voltage, amperage, electrode extension, and _____ _____, as well as other factors, dramatically affects the weld quality you produce.

3. The voltage is set on the welder, and the amperage is set by changing the _____ _____.

4. The higher the wire-feed speed, the higher the _____.

5. The welder must set both the welding _____ and wire-feed speed correctly to produce a satisfactory weld.

6. Because changes in the wire-feed speed automatically change the _____ _____, it is possible to set the amperage by using a chart and measuring the length of wire fed per minute.

7. The wire-speed control dial can be advanced or slowed to control the burn weld size and _____ rate.

8. The *gun angle, work angle,* and *travel angle* are names used to refer to the relation of the gun to the _____.

9. By manipulating the electrode travel angle for the flat and horizontal positions of welding to a 20° to 90° angle from the vertical, the _____ can be controlled.

10. Shallower angles are needed when welding thinner materials to prevent _____ _____.

11. Steeper, perpendicular angles are used for _____ materials.

12. Vertical up welds require a _____ gun angle.

13. A gun angle around _____° either slightly forehand or backhand works best for overhead welds.

14. The forehand technique is sometimes referred to as _____the weld bead, Figure 12-6A, and backhand may be referred to as pulling, or dragging, the weld bead, Figure 12-6C.

15. The term _____ is used when the gun angle is at approximately 90° to the work surface, Figure 12-6B.

16. The movement or weaving of the welding electrode can control the following characteristics of the weld bead: buildup, width, undercut, and _____ _____.

17. However, because moving the tip of the electrode off of the molten weld pool can cause a great deal of spatter, most GMA welding is done with a _____ _____ line movement.

18. The electrode control consists primarily of keeping its movement _____ _____ along the entire length of the joint.

19. You must maintain a _____ speed and weave along the joint as well as a constant welding gun angle and electrode extension.

20. The electrode extension (stickout) is the length from the contact tube to the _____ measured along the wire, Figure 12-9.

21. As the electrode extension is _____, there is a reduction in weld heat, penetration, and fusion, and an increase in buildup.

22. On the other hand, as the electrode extension length is _____, the weld heats up, penetrates more, and builds up less, Figure 12-11.

23. _____ welds are used when the strength of a full-length weld is not required, to speed up the welding process and reduce welding cost, to reduce welding heat distortion, or to help stop weld bead cracking, Figure 12-13.

24. Lap joints are made by overlapping the _____ of the two plates, Figure 12-22.

25. Because a lap joint may not be as strong in all directions, it is important to weld _____ sides of the joint when possible, Figure 12-29.

26. An outside corner joint is made by placing the plates at an angle to each other, with the edges forming a _____, Figure 12-30.

27. The tee joint is made by tack welding one piece of metal on another piece of metal at a _____ angle, Figure 12-36.

28. A welded tee joint can be strong if it is welded on _____ sides, even without having deep penetration, Figure 12-38.

CHAPTER 13

Flux Cored Arc Welding Equipment and Materials

Quiz

Name: _____

Date: _____

Class: _____

Instructor: _____

Grade: _____

Instructions: Carefully read Chapter 13 in the text and answer the following questions:

MULTIPLE-CHOICE

1. Flux cored arc welding can be done with or without
   a. a wire feeder
   b. additional flux
   c. shielding gas
   d. adjusting the voltage

2. In FCA welding molten droplets from the melted electrode travel across the arc and mix with the
   a. plate surface
   b. filler metal
   c. molten slag
   d. molten base metal

3. FCA welding done without a shielding gas is called
   a. self-shielding
   b. non shielded
   c. internal shielded
   d. basic shielded

4. The flux cored arc welding (FCAW) method that uses a shielding gas is called
   a. inner shield
   b. self-shielded
   c. dual-shielded
   d. gas shielded

5. On FCA welders as the electrode feed speed increases, the
   a. amperage decreases
   b. voltage increases
   c. voltage decreases
   d. amperage increases

6. With FCA welding, the welder sets the voltage, and as the weld is being made, the voltage
   a. increases
   b. remains constant
   c. decreases
   d. varies with travel speed

7. An advantage of flux cored arc welding is
   a. it can be done on any type of metal
   b. there is no postweld cleanup required
   c. little practice is required to make quality welds
   d. it has high deposition rates

8. A limitation of flux cored arc welding is that
   a. the slag is too hard to remove
   b. it is confined to ferrous metals and nickel-based alloys
   c. it can only be done in the flat and horizontal positions
   d. only metal ¼ in. thick and thicker can be welded

9. To prevent burnthrough on thinner materials the gun angle should be
   a. very shallow
   b. very steep
   c. perpendicular
   d. at a right angle to the weld joint

10. What can cause weld bead undercut?
    a. too short of an electrode extension
    b. too long of an electrode extension
    c. too fast a travel speed
    d. too low arc voltage

SENTENCE COMPLETION

In the space provided write the answer that completes the statement.

1. FCA welding makes a weld by having an arc between a continuously fed tubular wire electrode and the _____.

2. The molten weld metal is protected from contamination by the gases formed as the _____ of the wire electrode vaporizes.

3. One method of FCA welding done without a gas shield is called _____ _____.

4. The other method of FCA welding done with a shielding gas is called _____ _____.

5. The FCA welding equipment, setup, and operation are similar to that of the _____ _____ welding process.

6. On a CP welder that is used for FCA welding, the welding _____ _____ is set by the welder, and the voltage remains constant as the weld is made.

7. On FCA welders the amperage (current) is set by adjusting the _____ _____.

8. The major atmospheric contaminations come from _____, the major elements in air.

9. Flux additives improve _____ and other physical or corrosive properties of the finished weld.

10. The flux core additives that serve as deoxidizers, shielding gas formers, and slag-forming agents either protect the molten weld pool or help to remove _____ _____ from the base metal.

11. If oxygen in the air were to come in contact with the molten weld metal, the weld metal would quickly _____.

12. _____ helps the weld by protecting the hot metal from the effects of the atmosphere; controlling the bead shape by serving as a dam or mold; and serving as a blanket to slow the weld's cooling rate, which improves its physical properties.

13. The water-cooled FCA welding gun is more efficient at removing _____ _____ than an air-cooled gun.

14. _____ may be added to the water in recirculating systems to prevent freezing, to aid in pump lubrication, and to prevent algae growth.

15. Although small portable FCA welding machines have low input power requirements, they can make single-pass welds on metal as thick as 3/8 in. (10 mm) and multiple-pass welds on metal up to _____ and thicker.

16. FCA welding has no stub loss, so nearly _____% of the FCAW electrode purchased is used.

17. The addition of _____ and other fluxing agents permits high-quality welds to be made on plates with light surface oxides and mill scale.

18. Changes in power settings can permit welding to be made on thin-gauge sheet metals or thicker plates using the same _____.

19. The main limitation of flux cored arc welding is that it is confined to ferrous metals and _____.

20. The slight _____ or bend in the electrode wire ensures a positive electrical contact between the contact tube and filler wire.

21. The _____ or twist causes the electrode to twist as it feeds through the contact tip.

22. Slags can be refractory, become solid at a high temperature, and solidify over the weld, helping it hold its shape and slowing its _____.

23. Molten weld metal tends to have a high _____, which prevents it from flowing outward toward the edges of the weld.

24. Fluxing agents make the weld more _____ and allow it to flow outward, filling the undercut.

25. The addition of ferrite-forming elements can control the _____ and brittleness of a weld.

26. The welding fluxes form slags that are less dense than the weld metal so that they will float to the surface before the weld _____.

27. The AWS classification for stainless steel for FCAW electrodes starts with the letter _____ as its prefix.

28. Some wire electrodes require storage in an electric rod oven to prevent contamination from excessive _____.

29. To prevent hydrogen entrapment, porosity, and atmospheric contamination, it may be necessary to _____ the base metal to drive out moisture.

30. When the electrode provides all of the shielding, it is called _____.

31. Using a self-shielding flux cored electrode with a shielding gas may produce a _____ _____ weld.

32. The weld bead width, buildup, penetration, spatter, chemical composition, and mechanical properties are all affected as a result of the _____ _____ selection.

33. Gases used for FCA welding include CO_2 and mixtures of _____ _____ and CO_2.

34. The *gun angle, work angle,* and *travel angle* are terms used to refer to the relation of the gun to the _____.

35. Shallower angles are needed when welding thinner materials to prevent _____ _____.

36. Vertical up welds require a _____ gun angle.

37. _____ travel speeds deposit less filler metal.

38. The mode of metal transfer is used to describe how the molten weld metal is transferred across the _____ to the base metal.

39. The _____ transfer mode is the most common process used with gas-shielded FCAW.

40. The characteristic of _____ metal transfer is a smooth arc, through which hundreds of small droplets per second are transferred through the arc from the electrode to the weld pool.

41. To achieve a spray transfer, high current and _____ diameter electrode wire is needed.

42. Globular transfer occurs when the welding current is _____ _____ the transition current.

43. Both FCAW-S and FCAW-G use _____ when welding on thin-gauge materials to keep the heat in the base metal and the small diameter electrode at a controllable burn-off rate.

44. Larger diameter electrodes are welded with _____ because the larger diameters can keep up with the burn-off rates.

45. In the _____ welding position, the workpiece is placed flat on the work surface.

46. In the _____ welding position, the workpiece is positioned perpendicular to the workbench surface.

47. The electrode extension is measured from the end of the _____ _____ to the point the arc begins at the end of the electrode.

48. Compared to GMA welding, the electrode extension required for FCAW is much

_____.

49. Porosity can be caused by moisture in the flux, improper gun manipulation, or surface

_____.

50. New hot-rolled steel has a layer of dark gray or black iron oxide called _____

_____.

51. Often, it is better to remove the mill scale before welding rather than risking the production of _____.

52. All welding surfaces within the weld groove and the surrounding surfaces within 1 in.
(25 mm) must be cleaned to _____.

53. Any time FCA welds are to be made on metals that are dirty, oily, rusty, wet, or that
have been painted, the surface must be _____.

54. The most common causes of FCA welding problems are _____.

CHAPTER 14

Flux Cored Arc Welding

Quiz

Name: _____

Date: _____

Class: _____

Instructor: _____

Grade: _____

Instructions: Carefully read Chapter 14 in the text and answer the following questions:

MULTIPLE-CHOICE

1. The voltage on FCA welding machines is set by
   a. the welding machine manufacturer
   b. the equipment installer
   c. the voltage the equipment is connected to
   d. adjusting it on the welding machine

2. Using the wrong length of the electrode extension can result in
   a. better weld visibility and does not affect the weld
   b. too much weld smoke being produced
   c. slag that is very hard to remove
   d. a weld that has porosity and slag inclusions

3. If the wire-feed speed is too slow, what can you do to the FCA welding voltage to correct the problem?
   a. increase the voltage setting
   b. decrease the voltage setting
   c. change from DCEN to DCEP voltage
   d. change from DCEP to AC voltage

4. If the wire-feed speed is too high, then the voltage
   a. is too high
   b. is too low
   c. is set to the wrong polarity
   d. should be changed to AC

5. The best way to start the FCA welding setup is to set the voltage and wire
   a. at the lowest possible setting
   b. at the highest possible setting
   c. according to a welding setup table
   d. at the middle of the wire-speed and voltage setting

6. What type of weld is made through the surface of a plate directly below?
   a. plug weld
   b. edge weld
   c. surface weld
   d. back weld

7. What type of a weld can be used to join flanges on structural shapes like I-beams?
   a. plug weld
   b. surface weld
   c. edge weld
   d. back weld

8. What may happen to the edge of a thin section if it is overheated?
   a. it may burn back
   b. too much weld filler may be applied
   c. slag may be trapped in the root of the weld joint
   d. because FCAW filler metal is being automatically added to the joint, overheating cannot occur

9. What can a wedge and cleat be used for?
   a. holding the wire in place
   b. blocking the root opening
   c. bracing the plate in a vertical or out-of-position welding position
   d. aligning plates' edges

10. To prevent burn back on the thin end of a triangular piece the weld should
    a. start on the wider portion
    b. start in the middle and weld to both ends
    c. start on the narrowest portion
    d. use a back-stepping weld procedure

SENTENCE COMPLETION

In the space provided write the answer that completes the statement.

1. The more you practice and watch what happens during a weld, the better your welding _____ will become.

2. One of the major factors that affect FCA welding is that the welding wire is being fed out at a _____ speed.

3. The upside to the higher welding speeds is that FCA welding is highly productive and very _____.

4. The voltage is set on the welder, and the amperage is set by changing the _____ _____.

5. In other words, because the voltage remains constant in FCA welding, the _____ _____ rises automatically to match the faster electrode feed rate.

6. The welder must set both the welding _____ and wire-feed speed correctly to produce a satisfactory weld.

7. The wire-feed speed is generally recommended by the _____ and is selected in inches per minute (ipm), which can be measured by how fast the wire exits the contact tube.

8. Because at high wire-feed speeds many feet of wire can be fed out during a full minute's wire-feed speed test, a _____ time test is desirable.

9. Loose electrode wire can be a trip hazard as well as an electrical hazard if it comes in contact with the _____.

10. The wire-speed control dial can be advanced or slowed to control the burn weld size and _____.

11. The movement or weaving of the welding electrode can control the following characteristics of the weld bead: buildup, width, undercut, and _____.

12. Electrodes such as E70T-1 will allow you to weld over _____, but some adjustments in setup are needed.

13. Even with good welding power and work cables, some voltage can be lost, especially when the work cables are _____, so a higher voltage setting at the welder may be needed to compensate for this loss.

14. In addition to having the welder set correctly, you must control the _____ by keeping its movement consistent along the entire length of the joint.

15. Starting with the _____ position allows you to build your skills slowly so that out-of-position welds become easier to do.

16. The _____ (stickout) is the length from the contact tube to the arc measured along the wire.

17. Because the length of electrode extension affects the _____ of the electrode, charts for FCA welding setup usually include a specific electrode extension or a range for that extension.

18. Using the wrong electrode extension for a specific electrode can result in very poor welding with slag inclusions, porosity, and a _____.

19. The edge welds can be used to join thinner plate sections to make a _____ _____ section.

20. _____ welds are made by first cutting or drilling a hole through the top plate and making a weld through that hole onto the plate that is directly behind the top plate.

21. Sometimes plug welds are used to make _____ welds that would not show after the weldment is finished.

22. Some of the common problems that can occur if you do not use the _____ _____ pattern are that each piece gets a little larger in size or the straightness of the cuts decreases.

23. Welds should not start or stop in a _____.

24. Vertical down welding is often used to make welds on thin-gauge _____ _____.

25. One of the advantages of vertical down is that relatively high weld travel speeds are possible, and this helps reduce _____ and distortion.

CHAPTER 15

Gas Tungsten Arc Welding Equipment and Materials

Quiz

Name: _____

Date: _____

Class: _____

Instructor: _____

Grade: _____

Instructions: Carefully read Chapter 15 in the text and answer the following questions:

MULTIPLE-CHOICE

1. In gas tungsten arc welding, the electrode
   a. is consumable
   b. melts slowly to add to the filler metal
   c. does not melt
   d. must be kept as cool as possible

2. GTA welding is effective for welding material
   a. that is flat and not too large
   b. that is ferrous or ferrous based
   c. ranging from sheet metal up to 1/4 in.
   d. that can be chemically cleaned

3. Another name for the work clamp is
   a. shop clamp
   b. electrode clamp
   c. cable clamp
   d. ground clamp

4. An advantage of using an air-cooled GTA welding torch is
   a. it is cooler
   b. it is more portable
   c. air is free, so operating the torch is cheaper
   d. the weld is less likely to get bubbles of gas trapped causing porosity

5. Why is tungsten a good choice for the electrode in GTA welding?
   a. because it is very hard and can take the arc force
   b. because it can be ground to a sharp point
   c. because it is less expensive than other good conductors like silver
   d. because of its high melting temperature

6. Why are cerium, lanthanum, thorium, and zirconium added to tungsten electrodes?
   a. to make them harder
   b. so they can be sharpened better
   c. to improve their properties
   d. to improve the surface appearance of the welds

7. Why do some GTA welding machines have a preflow setting?
   a. so that the welder can get into position before the arc starts
   b. to allow the shielding gas to clear the air out of the welding zone
   c. it allows the tungsten to be cooled before the arc starts
   d. to pre-position the tungsten for welding

8. Which type of GTA welding current puts most of the heat on the workpiece?
   a. DCEP
   b. AC
   c. ACHF
   d. DCEN

9. What shielding gas can be added to argon to aid in weld penetration?
   a. helium
   b. hydrogen
   c. nitrogen
   d. carbon dioxide

10. Why does GTAW use a shielding gas?
   a. to protect the molten weld pool and the tungsten electrode from the harmful effects of air
   b. to increase the weld visibility
   c. so that the filler metal can be added to the weld pool
   d. to prevent arc blow

SENTENCE COMPLETION

In the space provided write the answer that completes the statement.

1. The gas tungsten arc welding (GTAW) process is sometimes referred to as _____ _____, or heliarc.

2. Under the correct welding conditions, the tungsten electrode does not melt and is considered to be _____.

3. To make a weld, either the edges of the metal must melt and flow together by themselves or _____ must be added directly into the molten pool.

4. To prevent oxidation from occurring, an inert gas flows out of the welding torch, surrounding the hot tungsten and molten weld metal shielding it from _____ _____.

5. GTA welding is efficient for welding metals ranging from sheet metal up to _____ _____ in.

6. Two of the advantages of GTA welding for welding fabrication are that it can be used to produce very high-quality welds and it can be used to weld on almost any _____ _____.

7. Four major components make up a GTA welding station; they are the welding power supply, often called the welder; the welding torch, often called a TIG torch; the work clamp, sometimes called the ground clamp; and the _____.

8. GTA welding torches are available water cooled or _____.

9. The advantages of _____ torches make them the preferred GTA welding torch type for most small shops.

10. The lower operating temperature and continuous operating ability of the _____ _____ GTA welding torches make them the preferred torch type for most production welding.

11. The amperage listed on a torch is the _____ rating and cannot be exceeded without possible damage to the torch.

12. Water-cooled torches have three hoses connecting to the welding machine. In addition to the _____ hose, they have two cooling water hoses.

13. _____ systems use water pumps to circulate water through the torch.

14. The nozzle or cup is used to direct the _____ directly on the welding zone.

15. Small nozzle diameters allow the welder to better see the molten weld pool and can be operated with _____ gas flow rates.

16. Larger nozzle diameters can give better gas coverage, even in _____ _____ places.

17. The high melting temperature and good electrical conductivity make _____ _____ the best choice for a nonconsumable electrode.

18. The tungsten inclusions are hard spots that cause stresses to concentrate, possibly resulting in weld _____.

19. _____ tungsten has the poorest heat resistance and electron emission characteristic of all the tungsten electrodes.

20. The improved electron emission of the thoriated tungsten allows it to carry approximately 20% more _____.

21. Zirconiated tungstens are more resistant to weld pool _____ than pure tungsten, thus providing excellent weld qualities with minimal contamination.

22. _____ is added to tungsten to improve the current-carrying capacity in the same manner as does thorium.

23. The type of finish on the tungsten must be specified as cleaned or _____ _____.

24. The _____ may be merely a flow regulator used on a manifold system, or it may be a combination flow and pressure regulator used on an individual cylinder.

25. High gas flow rates waste _____ and may lead to contamination.

26. The larger the nozzle size, the higher is the _____ rate permissible without causing turbulence.

27. _____ is the time during which gas flows to clear out any air in the nozzle or surrounding the weld zone.

28. The _____ is the time during which the gas continues flowing after the welding current has stopped.

29. The shielding gases used for the GTA welding process are argon (Ar), helium (He), hydrogen (H), _____, or a mixture of two or more of these gases.

30. The purpose of the shielding gas is to protect the molten weld pool and the tungsten electrode from the harmful effects of _____.

31. Because argon is denser than _____, it effectively shields welds in deep grooves in the flat position.

32. Helium offers the advantage of deeper _____.

33. Hydrogen additions are restricted to stainless steels because hydrogen is the primary cause of _____ in aluminum welds.

34. _____, which used to be called direct-current straight polarity (DCSP), concentrates about two-thirds of its welding heat on the work and the remaining one-third on the tungsten.

35. _____, which used to be called direct-current reverse polarity (DCRP), concentrates only one-third of the arc heat on the plate and two-thirds of the heat on the electrode.

36. _____ current is continuously switching back and forth between DCEN and DCEP.

37. The speed at which current changes back and forth is referred to by three different names that all mean the same thing—cycles, frequency, or _____ _____.

38. As the electrons leave the surface of the metal, they provide some surface _____ _____ or removal of these oxides.

39. The hot start allows a controlled surge of welding current as the arc is _____ _____ to establish a molten weld pool quickly.

40. Read and follow the manufacturer's instructions and safety guidelines any time you are _____ any equipment.

41. The desired end shape of a tungsten electrode can be obtained by grinding, breaking, remelting the end, or using _____.

42. A grinder is often used to clean a contaminated tungsten or to _____ _____ the end of a tungsten.

CHAPTER 16

Gas Tungsten Arc Welding

Quiz

Name: _____

Date: _____

Class: _____

Instructor: _____

Grade: _____

Instructions: Carefully read Chapter 16 in the text and answer the following questions:

MULTIPLE-CHOICE

1. A long tapered tungsten tip shape will produce
   a. welds with wider surfaces
   b. more slag inclusions and less root penetration
   c. welds that may be prone to centerline cracking
   d. narrow, deeply penetrating welds

2. A shallow, tapered tungsten tip shape will produce
   a. less stable arcs and more tungsten inclusions
   b. better welds in thick sections where less distortion is required
   c. wider, shallower penetrating welds
   d. faster starting arcs

3. Carbide precipitation occurs when
   a. carbon combines with chromium
   b. mild steel is welded at too high an amperage setting
   c. chromium combines with ferrite to form raindrop-like inclusions
   d. the metal is wet and droplets of water form on the back of the weld

4. Using as low a welding current setting as possible on some stainless steels can be used to
   a. reduce weld embrittlement
   b. stop weld discoloration
   c. prevent carbide precipitation
   d. save shielding gas and its cost

5. Starting the molten weld pool on aluminum can be hard because
   a. the tungsten does not get as hot when AC current is used
   b. its high electrical conductivity prevents the needed heat buildup from resistance
   c. a molten weld pool is difficult to form
   d. it has a high thermal conductivity

6. Which metal is known to have a high surface tension on its molten weld pool?
   a. steel
   b. aluminum
   c. stainless steel
   d. mild steel

7. Why is it important to learn how to strike the arc in the exact place it needs to be?
   a. because striking it outside of the weld area is considered a defect
   b. so that you can immediately add filler metal
   c. so that the weld bead can be started as soon as possible
   d. to prevent the tungsten from being contaminated by any surface contamination that was not cleaned off the metal surface

8. Moving the torch at a faster rate of travel will result in
   a. less tungsten contamination
   b. a smoother weld bead
   c. less surface oxidation
   d. narrower weld beads

9. Using a higher welding current will result in
   a. more weld reinforcement
   b. higher shielding gas usage
   c. a wider weld bead
   d. less tungsten contamination

10. Surface welds can be used to
    a. prevent weld distortion
    b. build up the surface of equipment that has worn down
    c. improve the appearance of a V-groove weld
    d. prevent oxidation of the weld

SENTENCE COMPLETION

In the space provided write the answer that completes the statement.

1. Low carbon and mild steel are two basic _____
   classifications.

2. The point on a tungsten is usually ground so that it is twice as long as the _____
   _____ of the tungsten.

3. Long, tapered points produce _____, deeply penetrat-
   ing welds; and shallow, tapered points produce wider, shallower, penetrating welds.

4. Some GTA welding filler metals have alloys in them called _____
   _____ that can help prevent porosity.

5. Other than the need for more _____ of the base metal
   and filler metal, the setup and manipulation techniques for stainless steel welds are
   the same as those for low carbon and mild steels.

6. The most common sign that there is a problem with a stainless steel weld is the bead
   _____ after the weld.

7. To reduce the possibility of carbide precipitation, use as low a welding current setting
   as possible and/or travel as _____ as possible along
   the joint to help reduce carbide precipitation.

8. The high thermal conductivity of aluminum rapidly pulls the heat away from the welding area, making it more difficult to form a _____.

9. The high _____ of molten aluminum makes it easier to
   control large molten weld pools.

10. Aluminum has a thin protective oxide layer at room temperature, but it forms a much
    thicker layer at welding temperatures if it is not protected from _____
    _____.

11. Both the base metal and the filler metal used in the GTAW process must be thoroughly
    _____ before welding.

12. Typical water-cooled GTA torches are used for heavy-duty production welding and
    _____ torches are used for lighter general jobs.

13. The process of starting an arc involves melting a small spot of metal and is referred to
    as _____.

14. The term _____ refers to the process of forming a
    molten weld pool and manipulating the GTA welding torch so that the molten weld
    pool is worked along the plate surface.

15. _____—refers to the speed that the torch is moved across the metal's surface.

16. Faster travel speeds result in narrower weld beads, and slower travel speeds result in _____ weld beads.

17. The higher the welding current, the _____ the weld bead, and the lower the welding current, the narrower the weld bead.

18. _____—refers to a side-to-side or circular pattern that the torch is moved in as the weld progresses along the metal surface.

19. The term *autogenous weld* refers to a fusion weld made without the addition of _____ _____.

20. Surfacing welds, sometimes called _____, should be uniform in width and reinforcement while having minimum penetration.

21. The _____ joint can be one of the more difficult joints to do with gas tungsten arc welding.

CHAPTER 17

Oxyfuel Welding and Cutting Equipment, Setup, and Operation

Quiz

Name: _____

Date: _____

Class: _____

Instructor: _____

Grade: _____

Instructions: Carefully read Chapter 17 in the text and answer the following questions:

MULTIPLE-CHOICE

1. What controls the working pressure?
   a. cylinder valves
   b. torch valves
   c. check valves
   d. regulators

2. The working pressure gauge shows
   a. the regulated pressure being controlled for the torch
   b. the pressure at the torch tip
   c. how much gas is remaining in the cylinder
   d. what the gas flow rate is

3. The cylinder pressure gauge shows
   a. how much gas is left in the cylinder
   b. the remaining cylinder pressure
   c. what type of gas is in the cylinder
   d. the flow rate of gas from the cylinder

4. Why do different gases have different types of fittings?
   a. so that you have to purchase adapters or new equipment
   b. to keep the regulators and other devices in proper alignment
   c. so that you will use the proper wrench to tighten the fittings correctly
   d. to prevent them from being connected to the wrong type of gas

5. Why would schools, automotive repair shops, and small welding shops use a combination-type torch?
   a. they are less expensive than straight cutting torches
   b. they come in smaller sizes
   c. they are more flexible
   d. they are easier for beginners to use

6. Why do systems need reverse flow valves?
   a. to prevent gases from accidentally flowing out of one hose through the torch body and then into the other hose
   b. to provide a spacer in the hoses
   c. to reduce the working pressure to a safe level
   d. as a way to stop a flashback

7. What will stop both reverse gas flow and the flame of a flashback?
   a. flashback arrestor
   b. a double-check valve
   c. higher gas pressures
   d. smaller diameter hoses

8. Each time the oxyfuel system is being shut down
   a. the cylinders should be returned to the storage room
   b. the regulator pressure-adjusting screw should be backed off
   c. the torch valves should be turned off as tight as possible
   d. the system should be leak checked

9. What should be used to find leaking valve seats?
   a. Teflon tape
   b. leak-detecting solution
   c. feel for leaks with the back of your hand because it is more sensitive than the palm of your hand
   d. an ultraviolet light

10. What is the first step in setting up an oxyfuel torch set?
    a. check for leaks
    b. attach the regulator to the cylinder
    c. safety chain the cylinders to the cart or wall
    d. set the working pressure

SENTENCE COMPLETION

In the space provided write the answer that completes the statement.

1. Oxyfuel processes are safe only when all of the proper _____ procedures have been followed.

2. All oxyfuel processes use a high-heat, high-temperature flame produced by burning a fuel gas mixed with _____.

3. The cylinder gas pressure must be reduced by using a _____ to lower working gas pressure.

4. _____ is the most widely used fuel gas because of its high temperature and concentrated flame, but about 25 other gases are available.

5. All pressure regulators _____ the high cylinder or system pressure to the proper lower working pressure.

6. A regulator works by holding the forces on both sides of a _____ _____ in balance.

7. Because some small single-stage pressure regulators have difficulty maintaining the proper working pressure under high flow rates, _____ regulators were developed.

8. The working pressure gauge shows the regulated pressure being controlled for the torch, and the _____ pressure gauge shows the remaining cylinder pressure.

9. The line drop is caused by the resistance of a _____ as it flows through a line.

10. The only way to accurately determine the amount of remaining gases in these cylinders is to _____ them.

11. Regulators may be equipped with either a _____ or a safety disk to prevent excessively high pressures from damaging the regulator.

12. If a rupture disk opens, it must then be _____ before the regulator can be used again.

13. The regulator pressure adjusting screw should be _____ each time the oxyfuel system is being shut down.

14. High-pressure valve seats that leak result in a _____ or rising pressure on the working side of the regulator.

15. Gauges that do not read "0" (zero) pressure when the pressure is released, or those that have a damaged glass or case, must be _____.

16. Both regulators and gas cylinder fittings have a variety of different designs for various types of gases to ensure that regulators cannot be _____ to the wrong gas or pressure.

17. The connections to the cylinder and to the hose must be kept free of _____ _____.

18. The outlet connection on a regulator is either a right-hand fitting for _____ _____ or a left-hand fitting for fuel gases.

19. There are no internal or external moving parts on a regulator or a gauge that require _____.

20. When welding is finished and the cylinders are turned off, the gas pressure must be _____ and the adjusting screw backed out.

21. The _____ hand torch is the most common type of oxyfuel gas torch used for welding and cutting.

22. The hand torch can be purchased as either a combination welding and cutting torch or a _____ torch only.

23. The larger the diameter of the hole in a welding tip, the _____ _____ the heating capacity.

24. The heating capacity of a tip determines the _____ range it can be used on.

25. The sizes of tip cleaners are given according to the _____ size of the hole they fit.

26. Metal-to-metal seal tips must be tightened with a _____.

27. Using the wrong method of tightening the tip fitting may result in _____ _____ to the torch body or the tip.

28. Dirty welding and cutting tips can be cleaned using a set of _____ _____ or tip drills.

29. A backfire occurs when a _____ goes out with a loud snap or pop.

30. A _____ occurs when the flame burns back inside the tip, torch, hose, or regulator.

31. The purpose of the reverse flow valve is to prevent gases from accidentally flowing out of one hose through the torch body and then into the _____.

32. A reverse flow of gas can occur through the torch if it is not turned off or the pressure is not properly _____.

33. A flashback arrestor will stop both reverse gas flow and the flame of a _____ _____.

34. The flashback arrestor is designed to quickly stop the flow of _____ _____ during a flashback.

35. Fuel gas hoses are _____ and have left-hand threaded fittings.

36. Oxygen hoses are _____ and have right-hand threaded fittings.

37. When hoses are not in use, the gas must be _____ and the pressure bled off.

38. A carburizing flame has an excess of _____.

39. A neutral flame has a balance of fuel gas and _____.

40. An oxidizing flame has an excess of _____.

41. A _____ lighter is the only safe device to use when lighting any torch.

CHAPTER 18
Oxyacetylene Welding

Quiz

Name: _____

Date: _____

Class: _____

Instructor: _____

Grade: _____

Instructions: Carefully read Chapter 18 in the text and answer the following questions:

MULTIPLE-CHOICE

1. An advantage of oxyacetylene welding is
   a. the cylinders are lightweight
   b. it is portable
   c. any type of metal can be welded
   d. it can be easily used to make long welds

2. What can happen to the tip of an oxyacetylene torch if the flame is turned down too low?
   a. the tip can overheat
   b. it will become too cold to make a weld
   c. the flame will go out
   d. it cannot be adjusted to the neutral flame setting

3. What effect does increasing the angle of the torch toward 90° have?
   a. it decreases the rate of heating
   b. it increases the rate of heating
   c. it causes the molten weld pool to be blown away
   d. it has no effect on welding but does cause problems with brazing

4. What effect does increasing the distance between the inner cone and the metal have?
   a. it causes the weld pool to flow out more
   b. it increases the rate of heating
   c. it decreases the rate of heating
   d. it causes the molten weld pool to glow redder

5. What effect can using a larger diameter welding rod have on making a weld on thin sheet metal?
   a. it can cool the molten weld pool
   b. it will cause a burn through
   c. large diameter welding rods cannot be used on thin sheet metal
   d. it will increase the weld penetration

6. What effect on the weld does the secondary flame have?
   a. it completes the burning so that the flame does not pollute the atmosphere
   b. it lights up the weld for better visibility
   c. it burns away the mill scale and rust in front of the weld
   d. it protects the molten weld pool from atmosphere contamination

7. Why should the end of a filler rod be bent?
   a. to make it easier to weld around corners
   b. for safety and easy identification
   c. so that your arm does not get tired holding the rod at an odd angle to weld
   d. it keeps the rods from rolling off of the welding table

8. A neutral oxygen and acetylene flame is
   a. when the oxygen and acetylene flows are balanced
   b. when the oxygen regulator is set at 5 psig and the acetylene is set at 10 psig
   c. when the oxygen regulator is set at 10 psig and the acetylene is set at 5 psig
   d. when there is only a little foam forming around the weld pool

9. Which oxyacetylene joint can be made with or without filler metal?
   a. butt joint
   b. lap joint
   c. tee joint
   d. outside corner joint

10. Which oxyacetylene joint is made by aligning the edges of two plates on a flat surface?
    a. outside corner joint
    b. lap joint
    c. butt joint
    d. tee joint

SENTENCE COMPLETION

In the space provided write the answer that completes the statement.

1. For small welding jobs and for hobbyists, gas welding is often the best welding method for thin metal and small parts because it is so easy to control the _____ _____.

2. An advantage of gas welding is that by changing the size of the tip you can change the welding _____.

3. Because a gas welding rig is so portable, it is frequently used to make these _____ _____ welds.

4. _____ steel is the easiest metal to gas weld.

5. The torch tip _____ should be used to control the weld bead width, penetration, and speed.

6. _____ is the depth into the base metal that the weld fusion or melting extends from the surface, excluding any reinforcement.

7. The correct _____ rate cools the tip and helps keep the flame heat off of the end of the tip.

8. Other factors that can be changed to control the weld size are the torch angle, the flame-to-metal distance, the welding rod size, and the way the torch is _____ _____.

9. The torch angle and the angle between the inner cone and the metal have a great effect on the speed of melting and size of the _____.

10. The distance between the inner cone and the metal ideally should be 1/8 in. to _____ in. (3 mm to _____ mm).

11. A larger welding _____ can be used to cool the molten weld pool, increase buildup, and reduce penetration.

12. The molten weld pool must be protected by the secondary flame to prevent the _____ _____ from contaminating the metal.

13. To prevent burnthrough, the torch should be raised or tilted, keeping the outer flame envelope over the molten weld pool until it _____.

14. An increase in sparks on clean metal means an _____ in weld temperature.

15. A decrease in sparks indicates a _____ in weld temperature.

16. When the sparks increase quickly, the torch should be pulled back to allow the metal to cool, preventing a _____.

17. During gas welding, sometimes the torch _____ might become out of adjustment as the result of slight changes in the regulator pressure, the tip become partially blocked by sparks, or other similar problems.

18. The more comfortable or relaxed you are, the easier it will be for you to make uniform _____.

19. It is important to feed the welding wire into the molten weld pool at a _____ _____ rate.

20. The end of the filler rod should have a hook bent in it so that you can readily tell which end may be _____ and so that the sharp end will not be a hazard to a welder who is working next to you.

21. Band saw cutting is easier than using a reciprocating-type saw because the metal can be more easily cut without excessive _____.

CHAPTER 19

Soldering, Brazing, and Braze Welding Processes

Quiz

Name: _____

Date: _____

Class: _____

Instructor: _____

Grade: _____

Instructions: Carefully read Chapter 19 in the text and answer the following questions:

MULTIPLE-CHOICE

1. Soldering and brazing are similar, but soldering is done at a temperature
   a. below 840°F
   b. above 840°F
   c. below 212°F
   d. above 212°F

2. An advantage of soldering and brazing is that
   a. they can be done on metal that cannot be cleaned
   b. the filler metal matches the base metal's color
   c. they are easy, so little or no practice is needed
   d. they can be both a permanent or temporary assembly method

3. The ability of a joint to withstand being pulled apart is
   a. torsion strength
   b. compression strength
   c. corrosion resistance
   d. tensile strength

4. The ability of a brazed joint to withstand a force parallel to the joint is the
   a. tearing strength
   b. ripping strength
   c. shear strength
   d. pulling strength

5. The ability of a joint to bend without failing is
   a. flexibility
   b. ductility
   c. deformability
   d. elasticity

6. Flux can be used with soldering and brazing to
   a. aid in the reduction of surface oils
   b. absorb surface finishes like paint
   c. remove a light surface oxide from the base metal surface
   d. keep the metal from changing colors

7. What is the difference between brazing and braze welding?
   a. brazing uses stronger alloys
   b. there is little or no capillary action on braze weld joints
   c. braze welding always uses brass rods and brazing uses bronze rods
   d. nothing, they are the same process

8. Which soldering process uses a molten flux bath?
   a. torch
   b. furnace
   c. dip
   d. induction

9. Which soldering process can add corrosion protection to the entire surface of the part?
   a. torch
   b. furnace
   c. dip
   d. induction

10. What is the paste range of a solder alloy?
    a. when both liquid and solid metals exist at the same time
    b. when the metal is smooth and can be applied as a paste to the part or joint
    c. when there is too much of one alloy and the mixture will not solidify into a solid
    d. it is the point where the alloy begins to soften from heat before any melting occurs

SENTENCE COMPLETION

In the space provided write the answer that completes the statement.

1. Soldering and brazing are _____ processes.

2. The only difference is that soldering is done at a lower temperature, below 840°F (449°C), and brazing is done at a _____ temperature, above 840°F (449°C).

3. In both processes, the filler metal is _____, and the base metal is heated.

4. At a high enough temperature the solder or braze metal flows out over the base metal surface and a strong bond is formed between the _____ and base metal.

5. Since the base metal does not have to melt, a _____ heat source can be used.

6. It is easy to join _____ metals, such as copper to steel, aluminum to brass, and cast iron to stainless steel.

7. Very thin parts or a thin part and a thick part can be joined without _____ _____ or overheating them.

8. Brazing is divided into two major categories named _____ and _____.

9. In the brazing process the parts that are being joined must be fitted together very _____.

10. This small spacing allows capillary action to draw the filler metal into the _____ _____ when the parts reach the proper phase temperature.

11. _____ action is the force that pulls water up into a paper towel or pulls a liquid into a very fine straw.

12. The parts joined by _____ welding may have a very open or loose fitting.

13. Braze welding uses more _____ than brazing.

14. The tensile strength of a joint is its ability to withstand being _____ _____.

15. As the joint spacing decreases, the surface tension increases the _____ _____ strength of the joint.

16. The shear strength of a brazed joint is its ability to withstand a force parallel to the _____.

17. For a solder or braze joint, the shear strength depends upon the amount of _____ _____ area of the base parts.

18. Ductility of a joint is its ability to _____ without failing.

19. The _____ resistance of a metal is its ability to be bent repeatedly without exceeding its elastic limit and without failure.

20. Corrosion resistance of a joint is its ability to resist _____ attack.

21. The compatibility of the base materials to the filler metal will determine the _____ _____ resistance.

22. Fluxes used in soldering and brazing have three major functions: they must remove any oxides that form as a result of heating the parts, they must promote wetting, and they should aid in _____ action.

23. The use of fluxes does not eliminate the need for good joint _____ _____.

24. As the parts are heated to the soldering or brazing temperature, the flux becomes more _____.

25. Fluxes that are active at _____ temperature must be neutralized (made inactive) or washed off after the job is complete.

26. Soldering and brazing methods are grouped according to the method with which heat is applied: torch, furnace, induction, dip, or _____.

27. The flame of the torch is one of the quickest ways of heating the material to be joined, especially on _____ sections.

28. When using a torch, it is easy to overheat or burn the parts, flux, or _____ _____.

29. The _____ method of heating uses a high-frequency electrical current to establish a corresponding current on the surface of the part.

30. Two types of dip soldering or brazing are used: molten flux bath and _____ _____.

31. With the _____ method, the soldering or brazing filler metal in a suitable form is preplaced in the joint, and the assembly is immersed in a bath of molten flux.

32. With the _____ method, the prefluxed parts are immersed in a bath of molten solder or braze metal, which is protected by a cover of molten flux.

33. As with all soldering or brazing operations, any movement of the parts as they cool from a liquid through the paste range to become a solid will result in _____ _____ in the filler metal.

34. The _____ method of heating uses an electric current that is passed through the part.

35. The _____ method uses high-frequency sound waves to produce the bond or to aid with heat in the bonding.

36. Soldering and brazing metals are _____—that is, a mixture of two or more metals.

37. A _____ range is the temperature range in which a metal is partly solid and partly liquid as it is heated or cooled.

38. Soldering alloys are usually identified by their major _____ elements.

39. _____ solders are most commonly used on electrical connections, air-conditioning and refrigeration drain piping, and for architectural accents where good corrosion resistance is needed.

40. _____ solders are used for plumbing because they are lead free and for refrigeration work.

41. The American Welding Society's classification system for brazing alloys uses the letter _____ to indicate that the alloy is to be used for brazing.

42. The spacing between the parts being joined greatly affects the _____ _____ strength of the finished part.

33. As with all soldering or brazing operations, any movement of the parts as they cool from a liquid through the paste range to become a solid will result in _____ in the filler metal.

34. The _____ method of heating uses an electric current that is passed through the part.

35. The _____ method uses high-frequency sound waves to produce the bond or to aid with heat in the bonding.

36. Soldering and brazing metals are _____. This is a mixture of two or more metals.

37. A _____ range is the temperature range in which a metal is partly solid and partly liquid as it is heated or cooled.

38. Soldering alloys are usually identified by their composition.

39. _____ solders are _____ electric _____ and electrical connections, as quality _____ where strong and secure connections where good _____ resistance is heated.

40. _____ solders are used for plumbing because they are _____ and for electrical work.

41. The Braze Welding _____ filler alloy _____ for brazing alloys are the _____ the filler alloy is _____ for more _____.

42. The spacing between the metals being joined greatly affects the _____ strength of the finished part.

CHAPTER 20

Soldering and Brazing

Quiz

Name: _____

Date: _____

Class: _____

Instructor: _____

Grade: _____

Instructions: Carefully read Chapter 20 in the text and answer the following questions:

MULTIPLE-CHOICE

1. Which of these metals can have alloys where one is a soldering alloy and another is a brazing alloy?
   a. silver
   b. steel
   c. copper
   d. nickel

2. To obtain a high solder joint strength the parts must
   a. be fitted with enough joint space so that as much alloy as possible can fit into the joint
   b. be fitted with very small joint spaces
   c. be grooved so that the alloy has something to grip onto
   d. solder cannot be used for any joint requiring high strength because it is too weak

3. The filler metal alloy BCuP is classified as
   a. a silver braze alloy
   b. a copper brazing alloy
   c. a phosphorus brazing alloy
   d. a soldering alloy

4. What is a practical application of surface brazing?
   a. joining different types of surface materials
   b. bonding a hard metal sheet over the softer metal surface of a mild steel plate
   c. building up worn or damaged surfaces
   d. there is no practical application since it is only used as a decoration

5. What is an indication that braze metal has been overheated?
   a. the braze metal turns bright red
   b. it will make a crackling sound as it burns
   c. black burned flux and white powdery zinc oxide
   d. the molten pool of braze metal will flow out uncontrollably over the surface

6. What is the name of the small flat spot on a drill bit that does little or no actual cutting?
   a. the blank center
   b. the dead center
   c. the dead spot
   d. the flat spot

7. Why must metal being drilled be clamped securely before drilling starts?
   a. so that the hole will be drilled square
   b. as a way to keep the bit cutting smoothly
   c. so that you can keep one hand free to push down smoothly on the drill press handle
   d. to prevent the metal from spinning if the bit grabs

8. Why should a part be center punched before drilling?
   a. so you do not forget where the hole should be drilled
   b. to prevent the drill bit from overheating
   c. to prevent the bit from chattering when aluminum is being drilled
   d. it makes starting the hole easier

9. Why should a tap be backed up ¼ turn after it has been turned ½ to ¾ turns?
   a. so that the tap does not overheat
   b. it allows the oil to flow into the cut
   c. it keeps the threads aligned so that they will be straight
   d. to allow the chips that are being cut to be broken free

10. What are the best types of flames to use for silver brazing?
    a. oxyacetylene or air MAPP®
    b. air hydrogen or oxyhydrogen
    c. air acetylene or air MAPP®
    d. a slightly oxidizing oxyacetylene flame

SENTENCE COMPLETION

In the space provided write the answer that completes the statement.

1. Brazing and soldering both fall under the same American Welding Society classification, and it is only the _____ required to melt the filler metals that separates soldering from brazing.

2. Soldering takes place at temperatures _____ 840°F (450°C), and brazing takes place at temperatures _____ 840°F (450°C).

3. One of the advantages of soldering and brazing is that they can make either a permanent or a _____ joint.

4. When done correctly, both soldering and brazing can produce joints that are several times stronger than the _____ itself.

5. To obtain this higher strength, the parts being joined must be fitted so that the joint spacing is very _____.

6. Braze welding does not need _____ action to pull filler metal into the joint.

7. You can control the heat for both brazing and soldering by raising and lowering the torch flame, increasing or decreasing the _____, and flashing the torch off or away from the metal.

8. Using prefluxed rods is easier for students than using _____ flux, which has to have the hot tip of the brazing rod dipped to apply.

9. Worn or damaged surfaces can be built up with _____.

10. The lower temperature used in brazing does not tend to create _____ _____ where the base metal becomes hardened as it often can in welding.

11. You do not have to remove the mill scale, the light oxide layer found on most hot-rolled steel, because the brazing flux will become active enough at the brazing temperature to _____ it.

12. Some of the signs that the braze metal is being overheated include the bead spreading out over the metal surface, the flux turning _____, and/or a white powdery zinc oxide forming alongside of the braze bead.

13. It is often easier to drill a small _____ first before trying to drill the larger diameter hole in metal.

14. Always securely clamp metal being drilled because if the drill bit grabs the metal, it will _____ around the bit.

15. Cadmium fumes can be hazardous, so you must avoid _____ these fumes during the brazing process.

16. Most solvents are flammable, so they should be used only in a _____ _____ area away from any source of ignition.

Oxyacetylene Cutting

Quiz

Name: _____

Date: _____

Class: _____

Instructor: _____

Grade: _____

Instructions: Carefully read Chapter 21 in the text and answer the following questions:

MULTIPLE-CHOICE

1. In oxyacetylene cutting the steel is cut by
   a. blowing away the molten metal
   b. vaporizing the metal
   c. rapid oxidation of the metal
   d. friction between the oxygen stream and the hot metal

2. The kindling temperature of steel in pure oxygen is approximately
   a. 1600°F to 1800°F (871°C to 982°C)
   b. 1000°F to 1500°F (538°C to 816°C)
   c. 840°F (449°C)
   d. 212°F (100°C)

3. Which fuel gas and oxygen neutral flame produces the highest temperature?
   a. MAPP®
   b. acetylene
   c. propane
   d. hydrogen

4. When safely setting up a cutting torch, the very first step is to
   a. secure the cylinders with a safety chain before doing anything else
   b. talk with a friend
   c. light the torch with a spark lighter
   d. set the pressures to 5 psig

5. Which metal can be easily cut with the OFC process?
   a. cast iron
   b. low carbon steel
   c. high nickel steel
   d. stainless steel

6. When first starting to cut, what fuel and oxygen pressures should be set?
   a. 5 psig fuel pressure and 10 psig oxygen pressure
   b. 10 psig fuel pressure and 35 psig oxygen pressure
   c. the pressure recommended by the manufacturer
   d. 15 psig fuel pressure and 60 psig oxygen pressure

7. Torches that use a tip mixing chamber are known as
   a. equal-pressure torches
   b. injector torches
   c. venturi torches
   d. low-pressure torches

8. Why are some cutting tips chrome plated?
   a. so they can be seen easier
   b. to prevent spatter from sticking to the tip
   c. to reflect the heat away from the tip
   d. because some welders like the look of a chrome-plated torch tip

9. The differences in the number of preheat holes in the cutting tip determine
   a. how high above the plate you want to set the torch tip before starting to cut
   b. how fast you want to cut
   c. the type of metal being cut—brass and copper require the most preheat holes
   d. the type of fuel gas to be used in the tip

10. The bottom edge of a good oxyacetylene cut should be
    a. rounded off
    b. melted and removed
    c. gouged
    d. square and slag-free

SENTENCE COMPLETION

In the space provided write the answer that completes the statement.

1. Oxyacetylene cutting (OFC-A) is the primary cutting process in a larger group called _____.

2. OFC is a group of oxygen cutting processes that uses heat from an oxygen fuel gas flame to raise the temperature of the metal to its _____ temperature.

3. When the metal is hot enough, a high-pressure stream of _____ _____ is directed onto the metal, causing it to be cut.

4. The processes in this group are identified by the type of _____ _____ mixed with oxygen to produce the preheat flame.

5. Because it can be used for both cutting and welding, _____ will remain the primary fuel gas.

6. A good oxyacetylene cut should not only be straight and square, but it also should require little or no postcut _____.

7. Oxyfuel gas cutting is used to cut _____ alloys.

8. Cutting goggles or cutting glasses are required any time you are using a _____ _____.

9. The _____ is the most common type of oxyfuel gas cutting torch used.

10. The combination welding and cutting torch offers more flexibility because a cutting head, welding tip, or heating tip can be attached quickly to the same _____ _____.

11. The added length of the dedicated cutting torch helps keep the operator farther away from the heat and sparks and allows _____ material to be cut.

12. _____ is mixed with the fuel gas to form a high-temperature preheating flame.

13. Two methods are used to mix the gases; one method uses a mixing chamber, and the other method uses an _____ chamber.

14. Torches that use a mixing chamber are known as equal-pressure torches because the gases must enter the mixing chamber under the same _____.

15. The injector works by passing the oxygen through a _____, which creates a low-pressure area that pulls the fuel gases in and mixes them together.

16. Torches with the tip slightly angled are easier for you to use when cutting _____ _____.

17. Torches with a right-angle tip are easier to use when cutting pipe, angle iron, I beams, or other _____.

18. Most cutting tips are made of copper alloy, but some tips are _____ _____.

19. The diameter, or size of the _____, determines the thickness of the metal that can be cut.

20. A larger diameter oxygen orifice is required for cutting _____ _____ metal.

21. To make it easier to select a tip, you can use a standard set of tip cleaners to find the size of the _____.

22. In all cases, start out with the pressure recommended by the particular _____ _____ of the equipment being used.

23. Check the _____ literature for tip size and type recommendations.

24. If the tip is clogged or dirty, clean the tip and clean out the orifices with the proper-size _____.

25. The amount of preheat flame required to make a perfect cut is determined by the type of _____ used and by the material thickness, shape, and surface condition.

26. The differences in the type or number of preheat holes determine the type of _____ _____ to be used in the tip.

27. If acetylene is used in a tip that was designed to be used with one of the other fuel gases, the tip may _____, causing a backfire or the tip to explode.

28. MPS gases are used in tips having eight preheat holes or in a two-piece tip that is not _____.

29. To check the assembled torch tip for a good seal, turn on the oxygen valve and spray the tip with a _____.

30. If the cutting tip seat or the torch head seat is damaged, it can be repaired by using a _____ designed for the specific torch tip and head, or it can be sent out for repair.

31. The setting up of a cutting torch system is exactly like setting up oxyfuel welding equipment except for the adjustment of gas _____.

32. The oxygen and acetylene cylinders must be securely chained to a cart or wall before the _____ are removed.

33. Cracking the cylinder valves is done to blow out any _____ that may be in the valves.

34. Before the cylinder valves are opened, back out the pressure regulating screws so that when the valves are opened the gauges will show _____ pounds working pressure.

35. If you are using a combination welding and cutting torch, the oxygen valve nearest the _____ must be opened before the flame adjusting valve or cutting lever will work.

36. Tip cleaners are small, round files; excessive use of them will greatly increase the _____.

37. Sometimes with large cutting tips, the tip will pop when the acetylene is turned off first; if that happens, turn the _____ off first.

38. If a piece of soapstone is used, it should be _____ properly to increase accuracy.

39. The scribe and punch can both be used to lay out an accurate line, but the _____ _____ line is easier to see when cutting.

40. On the back of most tip cleaning sets, the manufacturer lists the equivalent _____ size of each tip cleaner.

41. With a larger tip and a longer hose, the pressure must be set _____ _____.

42. The oxyfuel gas cutting torch works when the metal being cut rapidly _____ _____ or burns.

43. _____ point is the lowest temperature at which a material will burn.

44. If you must work around combustibles, wet the area first and keep a bucket of water and a _____ handy.

45. One way of steadying yourself is to lean against the _____ you are working on.

46. You must always know where the _____ from a cut are being thrown; make sure they are not being directed toward the fuel tanks, dead grass or straw, oil, or other combustible materials.

47. The cutting tip will catch small sparks and become dirty or _____
_____.

48. An excessive amount of preheat flame results in the top edge of the plate being
melted or _____.

49. If the cutting speed is too _____, the oxygen
stream may not have time to go completely through the metal, resulting in an
incomplete cut.

50. Too slow a cutting speed results in the cutting stream wandering, thus causing
_____ in the side of the cut.

51. A correct pressure setting results in the sides of the cut being _____
_____ and smooth.

52. A good cut should have a smooth, even sound, and the sparks should come off the
bottom of the metal more like a _____ than a spray.

53. _____ slag is very porous, brittle, and easily removed
from a cut.

54. _____ slag is attached solidly to the bottom edge of a
cut, and it requires a lot of chipping and grinding to be removed.

55. Slag is found on bad cuts, due to dirty tips, too much preheat, too slow a travel speed,
too short a coupling distance, or incorrect _____.

56. Most hand torches will not easily cut metal that is more than _____
_____ thick.

57. Poor-quality cuts require more time to _____ than is
needed to make the required adjustments to make a good weld.

58. Any piece being cut should be supported so that the torch flame will not cut through
the piece and into the _____.

59. Guides and supports allow the height and angle of the torch head to remain _____
_____.

60. Since the torch must be held in an exact position while making any accurate cut, you
will normally support the torch weight with your _____.

61. Starts and stops can be made better and more easily if one side of the metal being cut
is _____.

62. You can find out what the kerf width is for a torch by making a sample cut in scrap
metal and measuring its _____.

Plasma Arc Cutting

Quiz

Name: _____

Date: _____

Class: _____

Instructor: _____

Grade: _____

Instructions: Carefully read Chapter 22 in the text and answer the following questions:

MULTIPLE-CHOICE

1. Plasma cutting does not make the metal being cut very hot so
   a. it cannot cut thick sections
   b. there is less distortion
   c. it is harder to cut highly conductive metals like copper and aluminum
   d. the metal needs to be preheated if it is very thick

2. The electric current forms the plasma
   a. between the metal and the electrode
   b. in the plasma torch handle
   c. between the electrode tip and the nozzle tip
   d. in the gas stream between the tip and the work clamp

3. Compressed air can be supplied to a cutting torch by
   a. a compressor
   b. a gas cylinder
   c. a fan
   d. the venturi action of the torch tip

4. The speed that a plasma arc cuts is affected by material thickness and
   a. the type of plasma gas
   b. the type of plasma
   c. the nozzle size of the torch
   d. the power of the torch

5. The most common method of starting the arc is
   a. by touching the tungsten to the work
   b. with a pilot arc between the electrode tip and the nozzle tip
   c. with high frequency current to help the arc jump between the tungsten and the work
   d. by increasing the air flow until the arc starts

6. The closer the torch nozzle tip is to the work
   a. the more likely you are to miss the line you are cutting
   b. the easier it is to follow the line
   c. the lower the air pressure needed to form the plasma
   d. the narrower the kerf will be

7. What is dross?
   a. it is the slag from a plasma cut
   b. it is the molten metal that falls from a plasma cut
   c. it is the sparks flying out of the cut
   d. it is mostly unoxidized metal resolidified on the bottom of a cut

8. The most popular gas for plasma arc cutting is
   a. compressed air
   b. argon
   c. nitrogen
   d. oxygen

9. Which of the safety concerns for plasma arc cutting listed below has the greatest potential to be fatal?
   a. high noise level
   b. light radiation
   c. electrical shock
   d. hot sparks

10. What can be used with a plasma cutting system to reduce the noise level, fumes, and distortion?
    a. a strong vacuum system
    b. a blower
    c. a water table
    d. misters

SENTENCE COMPLETION

In the space provided write the answer that completes the statement.

1. A typical portable plasma cutting system can cut mild steel up to _____ in. thick.

2. Plasma cutters have the unique ability to cut metals without making them very _____.

3. Small plasma cutting machines can do many of the same cutting jobs that are done with an oxyacetylene torch, but without the expense of renting _____ _____.

4. Plasma machines can cut _____, including aluminum, stainless steel, and cast iron.

5. Plasma is created by an _____ and is an ionized gas that has both electrons and positive ions whose charges are nearly equal to each other.

6. The term _____ is the term most often used in the welding industry when referring to the arc plasma used in welding and cutting processes.

7. The plasma torch is a device that allows the creation and control of the _____ _____ for cutting processes.

8. The torch body is made of a special plastic that is resistant to high temperatures, ultraviolet light, and _____.

9. Torches are available with heads that are fixed at a 75° angle, 90° angle, or a 180° angle (straight), or they may have a _____ head that can be adjusted to any desired angle.

10. The cooling for low-powered torches is typically done by allowing the air to continue flowing for a short period of time after the cutting power has _____ _____.

11. On larger high-powered torches, cooling is typically provided by circulating _____ _____ through the head.

12. Most handheld torches have a manual power switch that is used to start and stop the power source, _____, and cooling water (if used).

13. The following are the parts that are most commonly replaced: electrode tip, nozzle insulator, nozzle tip, and _____.

14. The metal parts are usually made out of _____, and they may be plated.

15. Keeping the tip as _____ as possible lengthens the life of the tip and allows for better-quality cuts for a longer time.

16. The spacing between the electrode tip and the nozzle tip, called electrode _____ _____, is critical to the proper operation of the system.

17. The electrode setback space, between the electrode tip and the nozzle tip, is where the electric current forms the _____.

18. The nozzle, sometimes called the _____, is made out of high temperature-resistant material such as ceramic.

19. A number of power and control cables and gas and cooling water hoses may be used to connect the power supply with the _____.

20. The power cable must have a _____-rated insulation, and it is made of finely stranded copper wire to allow for maximum flexibility of the torch.

21. Most small shop plasma arc cutting torches use compressed _____ _____ to form the plasma and to make the cut.

22. Compressed air can be supplied by either an external compressor or an _____ _____ compressor.

23. The gas hose carries compressed air from the plasma machine to the _____ _____ and is made of a special plastic that is resistant to heat and ultraviolet light.

24. The control wire is a two-conductor, low-voltage, stranded copper wire that connects the power switch to the _____.

25. The production of the plasma requires a _____, high-voltage, constant-current (drooping arc voltage) power supply.

26. Some low-powered PAC torches will operate with as low as _____ amps of current flow.

27. The very high temperatures of the plasma process allow much _____ _____ traveling rates so that the same amount of heat input is spread over a much larger area.

28. A steel plate cut using the plasma process may have only a slight _____ _____ in temperature following the cut.

29. By being able to make most cuts without preheating, the plasma process greatly reduces fabrication _____.

30. You can make a thicker cut by traveling more slowly and, in some cases, weaving the torch back and forth to make a wider _____ so the cut can carry all the way through.

31. However, most manual plasma arc cutting speeds are around _____ in. per minute.

32. The most popular materials cut are carbon steel, stainless steel, aluminum, and _____.

33. Because the PAC process does not rely on the thermal conductivity between stacked parts, thin sheets can be _____ and cut efficiently.

34. _____ is the metal compound that resolidifies and attaches itself to the bottom of a cut.

35. It is possible to make cuts dross free if the PAC equipment is in good operating condition and the metal is not too _____ for the size of torch being used.

36. The standoff distance is the distance from the nozzle tip to the _____
_____.

37. Because the electrode tip is located inside the _____,
and a high initial resistance to current flow exists in the gas flow before the plasma is generated, it is necessary to have a specific starting method.

38. A _____ arc is an arc between the electrode tip and the nozzle tip within the torch head.

39. When the torch is brought close enough to the _____,
the primary arc will follow the pilot arc across the gap, and the main plasma is started.

40. The _____ is the space left in the metal as the metal is removed during a cut.

41. The closer the torch nozzle tip is to the work, the _____ the kerf will be.

42. Keeping the diameter of the nozzle orifice as small as possible will keep the _____
_____ smaller.

43. Too high a power setting will cause an increase in the kerf _____
_____.

44. As the travel speed is increased, the kerf width will decrease; however, the bevel on the sides and the dross formation will _____ if the speeds are excessive.

45. Because the sides of the plasma stream are not parallel as they leave the nozzle tip, there is a _____ left on the sides of all plasma cuts.

46. The most popular gas for PA cutting is compressed _____.

47. Too low a gas flow results in a cut having excessive _____
and sharply beveled sides.

48. Too high a gas flow produces a _____ cut because of turbulence in the plasma stream and waste gas.

49. The water table is used to reduce the _____ level, control the plasma light, trap the sparks, eliminate most of the fume hazard, and reduce distortion.

50. The chance that a fatal shock could be received from this equipment is much _____ _____ than for any other welding equipment.

51. Some type of ear protection is required to prevent _____ to the operator and other people in the area of the PAC equipment when it is in operation.

52. This process produces a large quantity of fumes that are potentially _____ _____.

53. The open circuit voltage on a plasma machine can be high enough to cause severe electrical shock or _____.

54. _____ cuts are the most common type of cuts made with PAC torches.

CHAPTER 23

Arc Cutting, Gouging, and Related Cutting Processes

Quiz

Instructions: Carefully read Chapter 23 in the text and answer the following questions:

MULTIPLE-CHOICE

1. A light that is a single-color wavelength and travels in a parallel wave is called a
   a. laser light
   b. system indicator light
   c. decorative light
   d. arc light

2. Some lasers for welding and cutting operate continuously and other lasers are
   a. spun
   b. pulsed
   c. bounced or reflected
   d. created in a vacuum

3. Popular gas lasers use nitrogen, helium, and
   a. carbon monoxide
   b. air
   c. carbon dioxide
   d. acetylene

4. Lasers are used for welding, cutting, and
   a. soldering
   b. brazing
   c. forming
   d. drilling

5. A characteristic of laser beam cutting is
   a. the part must be in the dark
   b. the part must be dark in color
   c. the part does not have to be electrically conductive
   d. the top surface must be smooth

6. Why do most laser beam drilling operations use a pulsed laser?
   a. because the cloud of vaporized material defuses the laser beam
   b. so that the operator can control the depth of the hole more accurately
   c. so that the laser machine does not overheat
   d. because a continuous laser beam would cut through the part and the table below

7. Most laser beam welding is performed on
   a. dark-colored metals
   b. thin materials
   c. 1 in. and thicker pieces
   d. bright shiny metals

8. An oxygen lance can be used to cut
   a. fine internal shapes
   b. concrete
   c. only the outside of small parts
   d. only thin sections of steel

9. A cutting process that does not put any heat into the material being cut is
   a. laser beam cutting
   b. water jet cutting
   c. plasma arc cutting
   d. pulsed laser beam cutting

10. What causes the cutting action when arc cutting electrodes are used?
    a. air blown through the center of the electrode
    b. air blown from the electrode holder
    c. oxygen blown through the center of the hollow electrode
    d. the jetting action created in the small cavity at the end of the electrode

SENTENCE COMPLETION

In the space provided write the answer that completes the statement.

1. A laser is an amplified form of light that is a single _____ wavelength that travels in parallel waves.

2. Most lasers used for cutting, drilling, and welding produce a laser light beam in the _____ range.

3. Manufacturers use the laser to do everything from burning information on products as small and hard as diamonds to guiding machines to grind, cut, punch, drill, and cut to accuracies within a few _____ of an inch.

4. The laser light, unlike ordinary light that spreads out, tends to remain in a very tight beam without _____ out.

5. Lasers can be divided into two major types: lasers using a solid material for the laser and lasers using a _____.

6. Each of these two types is divided into two groups based on their method of operation: lasers that operate continuously and lasers that are _____.

7. Today, the most popular industrial solid laser is the neodymium-doped yttrium aluminum garnet _____.

8. Popular gas-type lasers use nitrogen, _____, and carbon dioxide (CO_2) or mixtures of helium and nitrogen with CO_2.

9. A laser can be used to drill _____ (LBD) through the hardest materials, such as synthetic diamonds, tungsten carbide cutting tools, quartz, glass, or ceramics.

10. Lasers are used for welding (LBW) materials that are too thin or too _____ _____ to be welded with other heat sources.

11. The highly concentrated energy from a high-powered laser can cause the instantaneous melting or vaporization of the material being struck by the _____ _____.

12. All materials will _____ some of the laser's light, some more than others.

13. Laser beam cutting uses a high-pressure jet of _____ to blow the molten or vaporized material out the bottom of the cut.

14. Depending on the power rating and material thickness, some laser cutting machines can cut _____in. per minute in 1/8-in. steel.

15. Some laser cutting equipment can make _____ that are within ± a hundred thousandths of an inch, which is as accurate as some rough machining operations.

16. The part being cut does not have to be electrically _____, so materials like glass, quartz, wood, and plastic can be cut.

17. Nothing comes in contact with the part being cut except the _____.

18. The width of the _____ is very small, which allows the nesting of parts in close proximity to each other, which will reduce waste of expensive materials.

19. Most laser beam drilling operations use a _____ laser because the cloud of vaporized material caused by the laser will diffuse a continuous laser beam.

20. Most laser beam welding is performed on _____ materials.

21. Although the power of most lasers is relatively small, when compared to other welding processes, it is the laser's ability to concentrate the power into a _____ _____ area that makes it work so well.

22. Laser equipment is physically much _____ than most of the other welding or cutting power supplies.

23. The oxygen lance cutting process uses a _____ alloy tube.

24. The oxygen lance's unique method of cutting allows it to be used to cut material not normally cut using a _____ process.

25. The oxygen lances can be used to cut reinforced _____.

26. The oxygen lance is also used to cut _____ sections of cast iron, aluminum, and steel.

27. This operation produces both high levels of radiant heat and plumes of molten sparks and _____.

28. Water jet cutting is not a thermal cutting process; the cut is accomplished by the rapid erosion of the material by a high-pressure jet of _____.

29. Water jet cutting does not put any heat into the material being cut, and it is this lack of heat input that makes this process _____.

30. The addition of an _____ powder can speed up the cutting, allow harder materials to be cut, and improve the surface finish of a cut.

31. Because the arc cutting electrodes work with any standard _____ _____, this makes them the ideal choice for many cutting jobs.

32. Arc cutting electrodes differ from standard arc welding electrodes in that they burn back inside the outer flux covering, creating a small _____.

33. Arc cutting electrodes can be used to cut metal, pierce _____, or to gouge a groove for welding or gouge out a defective weld.

34. Unlike the oxyfuel process, the air carbon arc cutting process does not require that the base metal react with the cutting _____.

35. The ACA air stream blows the _____ away.

36. The copper coating helps decrease the carbon electrode overheating by increasing its ability to carry higher _____, and it improves the heat dissipation.

37. Most shielded metal arc welding power supplies can be used for _____ _____.

38. Air carbon arc cutting is most often used for _____ work.

39. Never cut on any material that might produce fumes that would be hazardous to your health without proper safety precautions, including adequate ventilation and/or the wearing of a _____.

40. Air carbon arc cutting can be used to remove a _____ from a part.

41. Arc gouging is the removal of a quantity of metal to form a groove or _____ _____.

42. Washing is the process sometimes used to remove large areas of _____ _____ so that hard surfacing can be applied.

43. The quantity and volume of sparks and molten metal spatter generated during this process are a major safety _____.

44. The _____ level is high enough to cause hearing damage if proper ear protection is not used.

45. If the used parts have paint, oils, or other contamination that might generate hazardous _____, they must be removed in an acceptable manner before any cutting begins.

46. U-grooving is used to remove defective welds and to prepare a thick metal joint so that a full _____ weld can be made.

Other Welding Processes

Quiz

Name: _____

Date: _____

Class: _____

Instructor: _____

Grade: _____

Instructions: Carefully read Chapter 24 in the text and answer the following questions:

MULTIPLE-CHOICE

1. What type of current is used in resistance welding?
   a. line voltage and high amperage
   b. low voltage and low amperage
   c. high voltage and low amperage
   d. very low voltage and high amperage

2. The welding process that joins similar and dissimilar metals by introducing high-frequency vibrations into the overlapping metals is
   a. friction welding
   b. upset welding
   c. ultrasonic welding
   d. percussion welding

3. In inertia welding, what generates the heat needed to soften the parts to be welded?
   a. friction
   b. high-frequency vibrations
   c. a chemical reaction
   d. a plasma

4. Laser beam welds can be as small as
   a. 0.01 in. (0.254 mm)
   b. 0.1 in. (2.54 mm)
   c. 0.001 in. (0.0254 mm)
   d. 1.0 in. (25.4 mm)

5. Plasma arc welding can be used to make single pass welds, with or without filler metal, in plates up to
   a. 1/4 in. (6 mm)
   b. 1/2 in. (13 mm)
   c. 3/4 in. (19 mm)
   d. 1.0 in. (25.4 mm)

6. In stud welding after the end of the stud and the underlying spot on the surface of the work have been properly heated
   a. gravity drops the stud into the molten spot on the plate's surface
   b. the stud and plate's surface melt together
   c. a small spot of filler metal fills the gap as the arc stops
   d. they are brought together under pressure

7. What type of wear does a part undergo when its surface is subjected to rubbing?
   a. impact
   b. abrasion
   c. corrosion
   d. erosion

8. Why is it a good thing when the steel around the tungsten carbide crystals wears away during digging?
   a. it allows some of the tungsten carbide crystals to drop out leaving sharp edges for easier digging
   b. it allows the hard tungsten carbide crystals to slide easier across the dirt
   c. it is a self-sharpening feature
   d. it makes it easier to resurface the part the next time

9. When using GMA, FCA, and GTA welding processes for hardfacing, care must be taken to avoid
   a. striking the arc on a good unworn surface
   b. putting on too much material
   c. postweld cracking
   d. dilution of the weld

10. A method of applying a coating of molten metal that does not heat the object up very much and is known as a "cold" method of building up metal is
   a. thermal spraying
   b. shielded metal arc hardfacing
   c. gas metal arc hardfacing
   d. plasma arc hardfacing

SENTENCE COMPLETION

In the space provided write the answer that completes the statement.

1. In the resistance welding process, the weld is made by clamping the parts together between the welding machine's electrodes; then an electric _____ _____ is passed through the parts to heat up the surfaces so that they will fuse together.

2. The parts are usually joined as a result of heat and pressure and are not simply _____ together.

3. The current for resistance welding is usually supplied by either a _____ _____ or a transformer/capacitor arrangement.

4. A _____, when used, stores the welding current until it is used.

5. _____ welding is the most common of the various resistance welding processes.

6. In this process, the weld is produced by the _____ obtained at the interface between the workpieces.

7. The welding time is controlled by a _____ built into the machine.

8. Ultrasonic welding is a process for joining similar and dissimilar metals by introducing high-frequency _____ into the overlapping metals in the area to be joined.

9. The temperature produced is below the melting point of the materials being joined, thus, no _____ occurs during the welding cycle.

10. Inertia welding is a form of _____ welding.

11. In laser beam welding, fusion is obtained by directing a highly concentrated beam of coherent _____ to a very small spot.

12. Because the heat is provided by a beam of light, there is no physical contact between the _____ and the welding equipment.

13. Since the laser beam is a _____ beam, it can operate in air or any transparent material, and the source of the beam need not even be close to the work.

14. Laser welds are _____, sometimes less than 0.001 in. (0.0254 mm).

15. A gas, or _____, is present in any electrical discharge if sufficient energy is present.

16. The two outstanding advantages of plasmas are higher _____ and better heat transfer to other objects.

17. In plasma arc welding, a plasma jet is produced by forcing _____ to flow along an arc restricted electromagnetically as it passes through a nozzle.

18. The plasma jet passing through the restraining orifice has an accelerated _____ _____.

19. Any known metal can be melted, even _____, by the plasma jet process, making it useful for many welding operations.

20. Stud welding is a semiautomatic or automatic arc welding process. An arc is drawn between a metal _____ and the surface to which it is to be joined.

21. *Hardfacing* is defined as the process of obtaining desired properties or dimensions by applying, using oxyfuel or arc welding, an integral layer of _____ _____ of one composition onto a surface, an edge, or the point of a base metal of another composition.

22. Hardfacing may involve building up surfaces that have become _____ _____.

23. Hardfacing metals are provided in the form of rods for oxyacetylene welding, electrodes for shielded metal arc welding, or in _____ form for automatic welding.

24. In hardfacing operations, oxyfuel welding permits the surfacing layer to be deposited by flowing molten filler metal into the _____.

25. In all types of surfacing operations, the metal should be cleaned of all loose scale, rust, dirt, and other foreign substances before the _____ is applied.

26. Hardfacing by arc welding may be accomplished by shielded metal arc, gas metal arc, gas tungsten arc, submerged arc, _____, or other processes.

27. The type of service to which a part is to be exposed governs the degree of _____ _____ required of the surfacing deposit.

28. Electrodes may be classified into the following three general groups: resistance to severe abrasion, resistance to both impact and moderate abrasion, and resistance to severe impact and moderately severe _____.

29. Care must be exercised when using the GMA, FCA, and GTA welding processes for hard-facing in order to avoid _____ of the weld.

30. Thermal spraying is the process of spraying molten metal onto a surface to form a _____.

31. Because the molten metal is accompanied by a blast of air, the object being sprayed does not heat up very much; therefore, thermal spraying is known as a "_____" method of building up metal.

CHAPTER 25

Welding Automation and Robotics

Quiz

Name: _____

Date: _____

Class: _____

Instructor: _____

Grade: _____

Instructions: Carefully read Chapter 25 in the text and answer the following questions:

MULTIPLE-CHOICE

1. A welding process that is completely performed by hand is a
   a. semiautomatic joining process
   b. machine joining process
   c. manual joining process
   d. automated joining process

2. What welding process is one in which the filler metal is fed into the weld automatically, and the welder controls most other functions?
   a. manual joining process
   b. semiautomatic joining process
   c. machine joining process
   d. automated joining process

3. In what welding process is the welding performed by equipment and the welder controls the welding progress by making adjustments as required?
   a. machine joining process
   b. manual joining process
   c. semiautomatic joining process
   d. automated joining process

4. What welding process is a dedicated process (designed to do only one type of welding on a specific part) that does not require the welder to make adjustments during the actual welding cycle?
   a. manual joining process
   b. semiautomatic joining process
   c. machine joining process
   d. automatic joining process

5. What automatic joining processes are similar to automatic joining except that they are flexible and more easily adjusted or changed?
   a. manual joining process
   b. semiautomatic joining process
   c. automated joining process
   d. machine joining process

6. The industrial robot is a main component in what welding process?
   a. manual joining process
   b. semiautomatic joining process
   c. machine joining process
   d. automated joining process

7. Because of the expense involved in special jigs and fixtures, automatic welding or brazing is best suited to
   a. small shops with low overhead cost
   b. prototype work
   c. college welding programs
   d. large-volume production runs

8. In the machine welding process, why would the welder need to make adjustments in travel speed or joint tracking?
   a. to ensure that the joint is being made according to specifications
   b. because the machines often get out of adjustment
   c. to correct for changes in the welding power supply or to realign the track when it is bumped
   d. no adjusting is ever needed once the machine is set up

9. Shielded metal arc welding (SMAW) is the most commonly used
   a. semiautomatic joining process
   b. manual welding process
   c. machine joining process
   d. automated joining process

10. The most commonly used semiautomatic arc welding processes are
   a. gas tungsten arc welding and flux cored arc welding
   b. shielded metal arc welding and torch brazing
   c. gas metal arc welding and flux cored arc welding
   d. stud welding and soldering

SENTENCE COMPLETION

In the space provided write the answer that completes the statement.

1. The first industrial robots were _____ robots that were used to move material with little repetitive accuracy required.

2. Automation has allowed manufacturers to increase productivity and cut costs, which makes their products more competitively _____.

3. A manual joining process is one that is completely performed by _____
   _____.

4. You control all of the manipulation, rate of _____,
   joint tracking, and, in some cases, the rate at which filler metal is added to the weld.

5. The rate of travel or speed at which the weld progresses along the joint affects the width, reinforcement, and _____ of the weld.

6. The rate at which filler metal is added to the weld affects the reinforcement, width, and _____ of the weld.

7. The most commonly used manual arc (MA) welding process is _____
   _____.

8. A semiautomatic joining process is one in which the _____
   is fed into the weld automatically.

9. The addition of filler metal to the weld by an automatic wire-feeder system enables you to increase the _____ of welds, productivity, and weld quality.

10. In the SMAW process, the electrode holder must be lowered steadily as the weld progresses to feed the electrode and maintain the correct _____.

11. The frequent stopping for rod and electrode changes, followed by restarting, wastes time and increases the number of weld _____.

12. In some welding procedures, each weld crater must be chipped and ground before the weld can be _____.

13. The most commonly used semiautomatic (SA) arc welding processes are gas metal arc welding (GMAW) and _____.

14. A machine joining process is one in which the welding is performed by equipment and you control the welding progress by making _____ as required.

15. Adjustments in travel speed, joint tracking, work-to-gun or work-to-torch distance, and current settings may be needed to ensure that the joint is made according to _____.

16. The work may move past a stationary welding or joining station, Figure 25-8, or it may be held _____ and the welding machine moves on a beam or track along the joint.

17. To minimize adjustments during machine welds, a _____ weld is often performed just before the actual weld is produced.

18. An automatic joining process is a _____ process (designed to do only one type of welding on a specific part) that does not require you to make adjustments during the actual welding cycle.

19. Automatic welding or brazing is best suited to _____ production runs because of the expense involved in special jigs and fixtures.

20. Automated joining processes are similar to automatic joining except that they are flexible and more easily adjusted or _____.

21. The industrial _____ is rapidly becoming the main component in automated welding or joining stations.

22. Industrial robots are primarily powered by electric stepping motors, hydraulics, or pneumatics and are controlled by a _____.

23. _____ can be used to perform a variety of industrial functions, including grinding, painting, assembling, machining, inspecting, flame cutting, product handling, and welding.

24. Most robots can perform movements in three basic directions: longitudinal (X), transverse (Y), and _____.

25. Parallel or multiple workstations increase the _____ cycle (the fraction of time during which welding or work is being done) and reduce cycle time (the period of time from starting one operation to starting another).

26. All personnel should be instructed in the _____ operation of the robot.

27. All personnel should be instructed in the location of an emergency power _____ _____.

28. _____ should be mounted around the floor and work area to stop all movement when unauthorized personnel are detected in the work area during the operation.

29. A signal should sound or flash before the robot starts _____.

30. Many welding applications can be improved by using automated equipment, thereby reducing welder _____, increasing productivity, and reducing weld defects.

31. As the number of companies using automated and robotic equipment has increased, the cost of equipment has significantly _____, and the versatility of the equipment has improved.

CHAPTER 26

Filler Metal Selection

Quiz

Name: _____

Date: _____

Class: _____

Instructor: _____

Grade: _____

Instructions: Carefully read Chapter 26 in the text and answer the following questions:

MULTIPLE-CHOICE

1. Manufacturers use numbering systems, trade names, color codes, or a combination of methods to identify
   a. filler metals
   b. stainless steels
   c. tool steels
   d. welding equipment setup

2. The most widely used numbering and lettering system is one developed by the
   a. American Society for Testing and Materials
   b. American Iron and Steel Institute
   c. National Electrical Manufacturers Association
   d. American Welding Society

3. Important welding electrode technical information provided by the manufacturer can include
   a. where the electrodes were made
   b. the number of welding electrodes per pound
   c. what color the electrodes are
   d. the percentage of flux to filler metal diameter

4. Technical information supplied by the manufacturer can help the welder decide
   a. who should do the welding
   b. how long the welds should be made before starting on a new piece
   c. where the welds should be made
   d. the welding conditions under which the electrode can be used

5. To serve as most of the filler metal in the finished weld and carry the welding current is a function of the
   a. tungsten electrode
   b. SMAW electrode core wire
   c. brazing rod
   d. flux

6. To provide some of the alloying elements and serve as an insulator is the function of the
   a. SMAW electrode flux covering
   b. welding cable insulation
   c. electrode holder
   d. electrode core wire

7. What is another name for high-temperature slags?
   a. refractory
   b. mold forming
   c. jetting
   d. spatter-free

8. The type of current, type of metal, thickness of the metal, and weld position are all
   a. important items to consider before you accept a welding job
   b. things that are listed on a welding electrode package
   c. factors that should be considered when selecting a filler metal
   d. relating to the skills needed to do a good welding job

9. Each AWS electrode classification has
   a. electrode diameters
   b. its own welding characteristics
   c. welding speed and appearance
   d. color-coding

10. In the AWS identification system for electrodes the E in the number E6012 means
    a. electrode
    b. electrical
    c. electronic
    d. efficient

SENTENCE COMPLETION

In the space provided write the answer that completes the statement.

1. Manufacturers may use their own numbering systems, trade names, color codes, or a combination of methods to identify _____.

2. The most widely used numbering and lettering system is the one developed by the

   _____.

3. Information that pertains directly to specific filler metals is readily available from most electrode _____.

4. The AWS publishes a variety of books, pamphlets, and charts showing the minimum _____ for filler metal groups.

5. General information given by manufacturers may include some or all of the following: welding electrode manipulation techniques, joint design, prewelding preparation, postwelding procedures, types of equipment that can be welded, welding currents, and

   _____.

6. Minimum _____ strength, psi—the load in pounds that would be required to break a section of sound weld that has a cross-sectional area of 1 sq in.

7. Yield point, psi—the point in low and medium carbon steels at which the metal begins to stretch when force (stress) is applied and after which it will not return to its _____ length.

8. Elongation, percent in 2 in. (51 mm)—the percentage that a 2-in. (51-mm) piece of weld will stretch before it _____.

9. Charpy V notch, foot-pound—the _____ load required to break a test piece of weld metal.

10. Chemical analysis of the weld deposit may also be included in the information given by _____.

11. Carbon (C)—As the percentage of carbon increases, the tensile strength increases, the hardness increases, and ductility is _____.

12. Copper (Cu)—As the percentage of copper increases, the corrosion resistance and cracking tendency _____.

13. Shielded metal arc welding electrodes, sometimes referred to as _____ _____, arc welding rods, stick electrodes, or simply electrodes, have two parts.

14. The functions of the core wire include the following: to carry the welding _____ _____ and to serve as most of the filler metal in the finished weld.

15. A _____ wire is the primary metal source for a weld.

16. The core wire also supports the coating that carries the fluxing and alloying materials to the _____ and weld pool.

17. Heat generated by the arc causes some constituents in the flux covering to decompose and others to _____, forming shielding gases.

18. These gases prevent the atmosphere from contaminating the weld metal as it transfers across the _____.

19. Welding fluxes can affect the penetration and contour of the _____.

20. Some high-temperature slags, called refractory, solidify before the weld metal solidifies, forming a mold that holds the _____ in place.

21. Different types of electrodes require different _____ settings even for the same-size welding electrode.

22. Some welding electrodes may be used to join more than one similar type of metal; other electrodes may be used to join together two different types of _____.

23. Some welding electrodes can be used to make welds in _____ positions.

24. The type of joint and whether it is grooved or not may affect the performance of the welding _____.

25. Some welding electrodes will build up faster, and others will _____ deeper.

26. Welding electrodes that will operate on low-amperage settings will have less heat input and cause less _____.

27. Postheating may be required to keep a weld zone from hardening or _____ when using some welding electrodes.

28. The general working conditions such as wind, dirt, cleanliness, dryness, and accessibility of the weld will affect the choice of welding _____.

29. Each AWS electrode classification has its own welding _____.

30. The characteristics of each manufacturer's filler metals can be compared to one another by using _____ supplied by the manufacturer.

31. The AWS classification system uses a series of letters and _____ in a code that gives the important information about the filler metal.

32. E—Indicates an arc welding _____.

33. B—Indicates a _____ filler metal.

34. Filler metals classified within these specifications are identified by a system that uses the letter E followed by a series of numbers to indicate the minimum _____ _____ strength of a good weld, the position(s) in which the electrode can be used, the type of flux coating, and the type(s) of welding current.

35. Arc blow, or arc _____, is the magnetic deflection of the arc from its normal path.

36. Proper filler metal selection is one of the most important factors affecting the successful welding of a _____.

37. If you are considering a large purchase of filler metals, it is advantageous for you to request _____ of the various filler metals from manufacturers so that you can test their performance in your applications.

CHAPTER 27
Welding Metallurgy

Quiz

Name: _____

Date: _____

Class: _____

Instructor: _____

Grade: _____

Instructions: Carefully read Chapter 27 in the text and answer the following questions:

MULTIPLE-CHOICE

1. Metals gain their desirable mechanical and chemical properties as a result of
   a. the way they are machined
   b. their alloying and heat treating
   c. how they are welded
   d. how thick they are

2. The quantity of thermal energy in matter is
   a. heat
   b. temperature
   c. both heat and temperature
   d. neither heat nor temperature

3. The level of thermal activity in matter is
   a. heat
   b. temperature
   c. both heat and temperature
   d. neither heat nor temperature

4. Heat that results in a change in temperature is called
   a. latent heat
   b. hidden heat
   c. sensible heat
   d. normal heat

5. The resistance of a metal to penetration is a property known as
   a. ductility
   b. elasticity
   c. tensile strength
   d. hardness

6. The property of a material that resists forces applied to pull metal apart is
   a. tensile strength
   b. ductility
   c. toughness
   d. compressive strength

7. Metals that do not stretch much before they break are
   a. ductile
   b. tough
   c. brittle
   d. compressive

8. What can be done to lower residual stresses and reduce cracking of welds?
   a. quenching the part
   b. grinding the part
   c. grooving the part
   d. preheating the part

9. The area where the weld heat has changed the metallurgical properties is called the
   a. heat-affected zone (HAZ)
   b. weld zone
   c. weld bead
   d. heat changed zone

10. The arrangement of atoms into three-dimensional patterns in a metal structure
    are called
    a. metal grain
    b. weld grains
    c. crystal lattice
    d. HAZ

SENTENCE COMPLETION

In the space provided write the answer that completes the statement.

1. You need to learn _____ to recognize that special attention might be needed when welding certain types of steel and to understand the kind of welding procedure that may be required.

2. Welding operations heat the metals, and that heating certainly changes not only the metal's initial structure but its _____ as well.

3. *Heat* and *temperature* are both terms used to describe the quantity and level of thermal _____ present.

4. Heat is the amount of _____ energy in matter.

5. One BTU is defined as the amount of heat required to raise 1 lb (0.45 kg) of _____ 1°F (17.2°C).

6. There are _____ forms of heat.

7. One is called _____ because as it changes, the change in temperature can be sensed or measured.

8. Latent heat is the heat required to change matter from one state to another, and it does not result in a _____ change.

9. When matter changes from a gaseous to a liquid state or from a liquid to a solid state, latent heat must be _____ .

10. Temperature is a measurement of the vibrating speed or frequency of the _____ _____ in matter.

11. The basic unit of measure is the _____ .

12. As matter becomes warmer, its atoms vibrate at a _____ frequency.

13. When the object becomes hot enough, the vibrating frequency of the atoms gives off visible _____ .

14. The mechanical properties of a metal can be described as those quantifiable properties that enable the metal to resist externally imposed forces without _____ _____ .

15. Probably the most outstanding property of metals is the ability to have their properties altered by some form of _____ treatment.

16. It is the responsibility of the _____ or engineer to select a metal that has the best group of properties for any specific job.

17. *Hardness* may be defined as resistance to _____.

18. Brittleness is the ease with which a metal will _____ or break apart without noticeable deformation.

19. Generally, as the hardness of a metal is increased, the brittleness is also _____ _____.

20. Ductility is the ability of a metal to be permanently twisted, drawn out, bent, or changed in shape without cracking or _____.

21. Toughness is the property that allows a metal to withstand forces, sudden shock, or bends without _____.

22. Toughness is measured most often with the _____ test.

23. Strength is the property of a metal to resist _____.

24. Tensile strength refers to the property of a material that resists forces applied to _____ metal apart.

25. Yield strength is the amount of strain needed to permanently _____ _____ a test specimen.

26. The yield point is the point during tensile loading when the metal stops _____ _____ and begins to be permanently made longer by deforming.

27. Ultimate strength is a measure of the load that _____ a specimen.

28. Metals that do not stretch much before they break are _____.

29. Compressive strength is the property of a material to resist being _____ _____.

30. Shear strength of a material is a measure of how well a part can withstand forces acting to cut or _____ it apart.

31. Torsional strength is the property of a material to withstand a _____ _____ force.

32. Strain is deformation caused by _____.

33. _____ is the ability of a material to return to its original form after removal of the load.

34. When the force on the material exceeds the elastic limit, the material will be _____ _____ permanently.

35. Impact strength is the ability of a metal to resist fracture under a _____ _____ load.

36. Solid matter is either crystalline or _____ in form.

37. Solids that are crystalline in form have an _____ arrangement of their atoms.

38. Amorphic materials have no _____ arrangement of their atoms into crystals.

39. The fundamental building blocks of all metals are atoms arranged in very precise three-dimensional patterns called crystal _____.

40. Some metals change their lattice structure when _____ above a specific temperature.

41. An alloy is a metal with one or more elements added to it, resulting in a significant change in the metal's _____.

42. It is inconvenient, if not impossible, to list all phases and temperatures at which alloys exist; that kind of information is summarized in graphs called _____ _____.

43. Metallurgy uses _____ letters to identify different crystal structures.

44. A eutectic composition is the lowest possible _____ temperature of an alloy.

45. While a metal is in the liquid-solid phase, it is very weak and any movement will cause _____ to form.

46. The addition of filler metal changes the alloy so that it is not as subject to hot _____ _____.

47. It is these changes in steel that allow it to be hardened and softened by heating, quenching, tempering, and _____.

48. Phase diagrams for other metal alloys are more complicated, but they are used in exactly the same way, they describe the effects of changes in _____ or alloying on different phases.

49. It is possible to replace some of the atoms in the _____ with atoms of another metal in a process called solid-solution hardening.

50. Quenching is the process of rapidly _____ a metal by one of several methods.

51. _____ is the process of reheating a part that has been hardened through heating and quenched.

52. Heating a carbon-iron alloy to a specific temperature and then quenching it rapidly in water can cause a crystalline structure called _____.

53. When metals are heated, _____ growth is expected.

54. To obtain a fine-grained structure, the metal must be heated quickly above the critical temperature and cooled quickly in a process called grain _____ _____.

55. Cold working is when metals are deformed at _____ temperature by cold rolling, drawing, or hammering, and the grains are flattened and elongated.

56. Preheat is used to reduce the rate at which welds _____.

57. Stress relieving consists of heating to a point below the lower transformation temperature where new grain growth would occur and holding it for a sufficiently long period to relieve locked-up stresses, then slowly _____.

58. Annealing, frequently referred to as full annealing, involves heating the structure of a metal to a high enough temperature to turn it completely _____.

59. _____ consists of heating steels to slightly above a specific temperature and holding for austenite to form, then followed by cooling (in still air).

60. The _____ temperature at which any such changes occur defines the outer extremity of the zone of change; that zone of change is called the heat-affected zone (HAZ).

61. Many welding problems and defects result from undesirable _____ _____ that can dissolve in the weld metal.

62. Hydrogen is the principal cause of _____ in aluminum welds and with GMAW welds on stainless steels.

63. Hydrogen problems are avoidable by keeping organic materials away from weld joints, keeping the welding consumables dry, and _____ the components to be welded.

64. Nitrogen comes from _____ drawn into the arc stream.

65. The primary problem with nitrogen is _____.

66. As with nitrogen, the common source of oxygen contamination, _____ _____, reaches the weld because of poor shielding or excessively long arcs.

67. Cold cracking is the result of hydrogen dissolving in the weld metal and then diffusing into the _____.

68. The hot cracks are caused by tearing the metal along partially fused grain boundaries of welds that have not completely _____.

69. Stainless steels rely on free chromium for their resistance to _____ _____.

Quiz

Name: _____

Date: _____

Class: _____

Instructor: _____

Grade: _____

Instructions: Carefully read Chapter 28 in the text and answer the following questions:

MULTIPLE-CHOICE

1. The ease with which a metal can be properly welded is known as
   a. ductility
   b. weldability
   c. hardness
   d. versatility

2. What can be done before starting a weld to reduce the stress caused by the weld and help the filler metal flow?
   a. grind the plate clean
   b. make a wider groove
   c. use a higher current setting
   d. preheat

3. Why would you use postheating on some welds?
   a. to control the formation of porosity
   b. so that any slag inclusions dissolve into the weld
   c. to prevent cracking of brittle metals
   d. as a way of preventing the weld from becoming too soft

4. The higher the carbon content of carbon steel
   a. the more difficult it is to weld
   b. the higher the welding current must be set
   c. the longer the arc length should be to prevent burning out the carbon
   d. the easier it is to weld

5. To produce a satisfactory weld on steel containing more than 0.40% carbon, what can be done?
   a. make the groove wider
   b. use a carbon-based flux
   c. preheat the metal
   d. add 10 more amps to the normal welding current for the electrode being used

6. Welding stresses occur because when metal is heated and cooled,
   a. the weld metals are stronger than the base metal
   b. it expands and contracts
   c. the weld metals are weaker than the base metal
   d. all welds have some flaws that cause stresses

7. Cast iron can crack if it is
   a. heated or cooled unevenly or too quickly
   b. too thick
   c. welded too hot
   d. too thin

8. When repairing a cast iron crack
   a. make one long weld from one end of the crack to the other
   b. start in the center and weld to the ends
   c. make large hot welds
   d. start welds on the ends of the crack and weld to the center

9. When welding aluminum, heat must be applied much faster to the weld area to bring the aluminum to the welding temperature because
   a. the bright surface reflects so much of the arc's heat
   b. welding too slowly on aluminum will cause it to crack
   c. of aluminum's high thermal conductivity
   d. of aluminum's electrical conductivity

10. Aluminum welds shrink about ____ in volume as the weld solidifies, which can cause joint distortion.
    a. 2%
    b. 6%
    c. 13%
    d. 45%

SENTENCE COMPLETION

In the space provided write the answer that completes the statement.

1. The term _____ has been coined to describe the ease with which a metal can be welded properly.

2. _____ weldability means that almost any process can be used to produce acceptable welds and that little effort is needed to control the procedures.

3. _____ weldability means that the processes used are limited and that the preparation of the joint and the procedure used to fabricate it must be controlled very carefully or the weldment will not function as intended.

4. Parts break because they are worn out, damaged in an accident, underdesigned for the work, or _____.

5. If the wrong filler metal is selected, the weld can have major defects and not be fit for _____.

6. _____ the part before starting the weld reduces the stress caused by the weld and helps the filler metal flow.

7. Postheating slows the cooldown rate following welding, which prevents postweld _____ of brittle metals.

8. When large welds are needed, it is better to make more _____ welds than a few large welds.

9. Two primary numbering systems have been developed to classify the standard construction grades of _____, including both carbon and alloy steels.

10. One classification system was developed by the Society of Automotive Engineers (SAE); the other system is sponsored by the _____.

11. In both steel classification systems, the first number often, but not always, refers to the basic type of _____.

12. The first two digits together indicate the _____ within the basic alloy group.

13. The last two or three digits refer to the approximate permissible range of _____ content.

14. Steels alloyed with carbon and only a low concentration of silicon and manganese are known as _____ carbon steels.

15. _____ steels contain specified larger proportions of alloying elements.

16. All carbon steels can be welded by at least one method; however, the higher the carbon content of the metal, the more _____ it is to weld the steel.

17. _____ carbon steel has a carbon content of 0.15% or less, and mild steel has a carbon content range of 0.15% to 0.30%.

18. The welding of _____ carbon steels, having 0.30% to 0.50% carbon content, is best accomplished by the various fusion processes, depending upon the carbon content of the base metal.

19. For steels containing more than 0.40% carbon, preheating and subsequent heat treatment generally are required to produce a _____ weld.

20. _____ carbon steels usually have a carbon content of 0.50% to 0.90%.

21. These steels are much more _____ to weld than either the low or medium carbon steels.

22. Stainless steels consist of four groups of alloys: austenitic, ferritic, martensitic, _____ _____.

23. The most widely used stainless steels are the chromium-nickel _____ _____ types.

24. Keeping the carbon content low in stainless steel will also help reduce _____ _____.

25. If the carbon content of the metal is very _____, little chromium carbide can form.

26. Some filler metals have stabilizing elements added to prevent _____ _____.

27. The closer the characteristics of the deposited metal match those of the material being welded, the better is the _____ resistance of the welded joint.

28. The diameter of the electrode used to weld steel that is thinner than 3/16 in. (4.8 mm) should be equal to, or slightly less than, the _____ of the metal to be welded.

29. To weld stainless steels, the arc should be as _____ as possible.

30. Gray cast iron is easily welded but because it is somewhat porous it can absorb oils into the surface, which must be _____ before welding.

31. Almost all grades of alloy cast iron can be easily welded if care is taken to slowly preheat and _____ the part to prevent changes in the carbon and iron structure.

32. The major purpose of preheating and _____ of cast iron is to control the rate of _____ change.

33. Welding stresses occur because as metal is heated and cooled, it expands and _____ _____.

34. The faster an iron-carbon metal is quenched, the harder, more brittle, less ductile, and higher in _____ the metal will become.

35. The slow cooling of an iron-carbon metal from a high temperature is called annealing.

36. Cracks in parts that cannot be preheated to the desired level can still be welded, but the welds will be very hard and are more likely to _____.

37. Back-stepping welds are short welds that start ahead of the ending point of the first weld and go back to the end of the first _____.

38. One of the characteristics of aluminum and its alloys is that it has a great affinity for _____.

39. Aluminum can be arc welded using _____ welding rods.

40. Aluminum and its alloys can rapidly conduct _____ away from the weld area.

41. When aluminum welds solidify from the molten state, they will shrink about _____% in volume.

42. Repair, or _____, welding is one of the most difficult types of welding.

43. The part is often dirty, oily, and painted, and it must be _____ _____ before welding.

44. Contamination can be removed by sandblasting, grinding, or using _____ _____.

45. Before the joint can be prepared for welding, you must try to identify the _____ _____ of metal.

46. It takes a skilled welder with an understanding of all the various characteristics of the metals and types of welding to fix worn or _____ parts.

CHAPTER 29
Welder Certification

Quiz

Name: _____
Date: _____
Class: _____
Instructor: _____
Grade: _____

Instructions: Carefully read Chapter 29 in the text and answer the following questions:

MULTIPLE-CHOICE

1. The difference between being a qualified and certified welder is
   a. the certified welder has been a qualified welder for more than a year
   b. the qualified welder has his or her test plates to prove that he or she can weld
   c. the certified welder has his or her test plates to prove that she or he can weld
   d. the certified welder has a written report

2. The entry-level welder certification was developed by
   a. AISI
   b. AWS
   c. API
   d. ASME

3. Once you pass a certification test it means
   a. you can make only welds within the limitations of that certification
   b. you can make any type of weld with any process
   c. you can apply for a job and never have to take a weld test again
   d. you are a highly skilled welder capable of welding anything

4. A change in the joint geometry can require that a welder
   a. get written approval from the shop foreman before welding
   b. practice making the new welds for at least a day
   c. be retested
   d. must explain to his or her boss how the joint differs and how he or she would make the weld

5. What is the name of the document that lists the very specific prescribed standards a weld was made to?
   a. Prequalified Weld Test Procedure
   b. Qualification Test Record
   c. AWS D-1.1 for Structural Steel
   d. Procedure Qualification

6. One of the requirements for a student to pass the entry-level welder certification is to have a test grade of _____ on all areas except safety, and on the safety test you need a grade of _____.
   a. 70% and 80%
   b. 85% and 95%
   c. 75% and 90%
   d. 65% and 85%

7. Once you have passed the knowledge test for the entry-level welder certification, what must you do to be certified?
   a. pass two bend tests
   b. weld for more than one week
   c. show the teacher that you can pass the visual weld inspection test
   d. fill out the AWS forms

8. The acceptable undercut criteria for a face- and root-bend specimen is
   a. 5% of the base metal or 1/8 in. (3.2 mm), whichever is less
   b. 3/32 in. (2.4 mm)
   c. 10% of the base metal or 1/32 in. (0.8 mm), whichever is less
   d. no undercut is allowed on test plates

9. Once the specimens have been bent, no defect can be larger than
   a. 1/32 in. (0.8 mm)
   b. 10% of the specimen's thickness
   c. 3/8 in. (9.5 mm)
   d. 1/8 in. (3.2 mm)

10. The root gap on a test plate with a backing strip should be approximately
    a. 1/8 in. (3.2 mm)
    b. 1/4 in. (6.4 mm)
    c. 1/32 in. (0.8 mm)
    d. 0 in. (0 mm)

SENTENCE COMPLETION

In the space provided write the answer that completes the statement.

1. Welding, in most cases, is one of the few professions that requires job applicants to demonstrate their skills even if they are already _____.

2. A method commonly used to test a welder's ability is the qualification or _____ _____ test.

3. Welders who have passed such a test are referred to as _____ welders; if proper written records are kept of the test results, they are referred to as certified welders.

4. Being certified does not mean that a welder can weld everything, nor does it mean that every weld that is made is _____.

5. To ensure that a welder is consistently making welds that meet the standard, welds are inspected and _____.

6. Most qualifications and certifications are restricted to a single welding process, position, metal, and _____ range.

7. Welders can be certified in each welding process such as SMAW, GMAW, FCAW, GTAW, EBW, and _____.

8. The type of metal—such as steel, aluminum, stainless steel, and titanium—being welded will require a change in the _____.

9. Each certification is valid on a specific range of _____ of base metal.

10. Changes in the classification and size of the filler metal can require _____ _____.

11. If the process requires a _____ gas, then changes in gas type or mixture can affect the certification.

12. In most cases, a weld test taken in the flat position would limit certification to flat and possibly _____ welding.

13. Changes in weld type such as _____ or fillet welds require a new certification.

14. Welder performance qualification is the demonstration of a welder's ability to produce welds meeting very specific, prescribed _____.

15. The form used to document this test is called the _____.

16. A _____ document is the written verification that a welder has produced welds meeting a prescribed standard of welder performance.

17. The AWS entry-level welder qualification and certification program specifies a number of _____ not normally found in the traditional welder qualification and certification process.

18. A written test must be passed with a minimum grade of _____% on all areas except safety.

19. The safety questions must be answered with a minimum accuracy of _____%.

20. If you have passed the knowledge and safety test and successfully pass the two bend tests and/or any one of the several workmanship assembly weldments, you can be _____.

21. The weld specimen must first pass visual inspection before it can be prepared for _____ testing.

22. The weld reinforcement and backing strip, if used, must be _____ _____ flush to the surface.

23. All corners must be rounded to a radius of 1/8 in. (3.2 mm) maximum, and all grinding or machining marks must run _____ on the specimen.

24. The weld must pass both the face and _____ bends to be acceptable.

25. On all but short welds, the welding bead will need to be restarted after a welder stops to change _____.

26. When a weld bead is nearing completion, it should be tapered so that when it is _____ _____ the buildup will be more uniform.

27. The slag should always be chipped and the weld crater should be _____ _____ each time before restarting the weld.

28. The _____ should be restarted in the joint ahead of the weld.

29. The movement to the root of the weld and back up on the bead serves both to build up the weld and reheat the metal so that the depth of _____ will remain the same.

CHAPTER 30

Testing and Inspecting Welds

Quiz

Name: _____
Date: _____
Class: _____
Instructor: _____
Grade: _____

Instructions: Carefully read Chapter 30 in the text and answer the following questions:

MULTIPLE-CHOICE

1. What determines the extent to which a product is subjected to inspection?
   a. how much the part cost
   b. how good the welders are that built the product
   c. who welded it
   d. its intended service

2. The mechanical testing method of quality control
   a. requires that more than two parts be tested for each 100 products
   b. requires that the welder present his or her certification papers before testing can begin
   c. results in the product being destroyed
   d. is only used on small GTA welded products

3. In what testing method can the part be used for its intended purpose after the test?
   a. guided-bend testing
   b. nondestructive testing
   c. tensile testing
   d. mechanical testing

4. When a discontinuity becomes so large that the weld is not acceptable under the standards for the code for that product, it becomes
   a. a flaw
   b. subject to more testing
   c. a defect
   d. de-rated to a lower standard

5. A discontinuity that results from gas that is dissolved in the molten weld pool, forming bubbles that are trapped as the metal cools to become solid is called
   a. porosity
   b. slag inclusions
   c. metallic inclusion
   d. cold lap

6. What are nonmetallic materials, such as slag and oxides, which are trapped in the weld metal called?
   a. porosity
   b. overlap
   c. cold lap
   d. inclusions

7. The lack of coalescence between the molten filler metal and previously deposited filler metal and/or the base metal is called
   a. slag inclusions
   b. incomplete fusion
   c. metallic inclusion
   d. porosity

8. A welding problem that a welder cannot control that can occur as the result of internal plate defects is
   a. lamellar tears
   b. overlap
   c. metallic inclusion
   d. cold lap

9. What type of testing is used to determine how well a weld can resist repeated fluctuating stresses or cyclic loading?
   a. tensile testing
   b. magnetic particle testing
   c. impact testing
   d. fatigue testing

10. Which weld test uses one or more hammer blows?
    a. free-bend test
    b. fatigue test
    c. nick-break test
    d. tensile test

SENTENCE COMPLETION

In the space provided write the answer that completes the statement.

1. Items that are to be used in light, routine-type service, such as ornamental iron, fence posts, gates, and so forth, are not inspected as _____ as products in critical use.

2. The two classifications of methods used in product quality control are destructive, or mechanical, testing and _____ testing.

3. Mechanical testing (DT) methods, except for hydrostatic testing, result in the product being _____.

4. Nondestructive testing (NDT) does not _____ the part being tested.

5. Nondestructive testing is used for weld qualification, welding procedure qualification, and product _____ control.

6. A discontinuity becomes a defect when the discontinuity becomes so large or when there are so many small _____ that the weld is not acceptable under the standards for the code for that product.

7. The difference between what is acceptable, fit for service, and perfection is known as _____.

8. When evaluating a weld, it is important to note the type of discontinuity, the size of the discontinuity, and the _____ of the discontinuity.

9. Porosity results from gas that was dissolved in the molten weld pool, forming _____ _____ that are trapped as the metal cools to become solid.

10. Porosity is most often caused by improper welding techniques, contamination, or an improper _____ balance between the filler and base metals.

11. The intense heat of the weld can decompose paint, dirt, or oil from machining and rust or other oxides, producing _____.

12. Uniformly scattered porosity is most frequently caused by poor welding techniques or faulty _____.

13. Clustered porosity is most often caused by improper starting and _____ _____ techniques.

14. Linear porosity is most frequently caused by _____ within the joint, root, or interbead boundaries.

15. Piping porosity, or wormhole, is most often caused by contamination at the _____ _____.

16. Inclusions are nonmetallic materials, such as slag and oxides, that are trapped in the weld metal, between weld beads, or between the weld and the _____ _____ metal.

17. Scattered inclusions can resemble porosity but, unlike porosity, they are generally not _____.

18. Inadequate joint penetration occurs when the depth that the weld penetrates the _____ is less than that needed to fuse through the plate or into the preceding weld.

19. Incomplete fusion is the lack of coalescence between the molten filler metal and previously deposited filler metal and/or the _____.

20. Arc strikes, even when ground flush for a guided bend, can open up to form small cracks or _____.

21. Overlap is also called _____, and it occurs in fusion welds when weld deposits are larger than the joint is conditioned to accept.

22. Undercut is the result of the _____ force removing metal from a joint face which is not replaced by weld metal.

23. To prevent undercutting, you can weld in the flat position by using multiple instead of _____ passes, change the shield gas, and improve manipulative techniques to fill the removed base metal along the toe of the weld bead.

24. Crater cracks are the tiny cracks that develop in the weld _____ _____ as the weld pool shrinks and solidifies.

25. Underfill on a groove weld appears when the weld metal deposited is inadequate to bring the weld's face or root surfaces to a level equal to that of the _____ _____ plane or plate surface.

26. Some problems result from internal plate defects that you cannot _____ _____.

27. Lamellar tears appear as cracks parallel to and under the steel _____ _____.

28. Laminations differ from lamellar tearing because they are more extensive and involve _____ layers of nonmetallic contaminants.

29. When laminations intersect a joint being welded, the heat and stresses of the weld may cause some laminations to become _____.

30. Tensile tests are performed with specimens prepared as round bars or flat _____ _____.

31. After the weld section is machined to the specified dimensions, it is placed in the tensile testing machine and pulled _____.

32. Fatigue testing is used to determine how well a weld can resist repeated fluctuating _____ or cyclic loading.

33. The two forms of shearing strength of welds are transverse shearing strength and longitudinal _____ strength.

34. The three methods of testing welded butt joints are (1) the nick-break test, (2) the guided-bend test, and (3) the _____ test.

35. To test welded, grooved butt joints on metal that is 3/8 in. (10 mm) thick or less, two specimens are prepared and tested—one face bend and one _____ _____ bend.

36. The free-bend test is used to test welded joints in _____.

37. One common test is the Izod test, Figure 30-34A, in which a notched specimen is struck by an anvil mounted on a _____.

38. Nondestructive testing of welds is a method used to test materials for _____ _____ defects such as cracks, arc strikes, undercuts, and lack of penetration.

39. Visual inspection is the most frequently used _____ testing method and is the first step in almost every other inspection process.

40. Visual inspection should be used before any other nondestructive or mechanical tests are used to _____ (reject) the obvious problem welds.

41. Penetrant inspection is used to locate minute surface cracks and _____ _____.

42. Magnetic particle inspection uses finely divided ferromagnetic particles (powder) to indicate defects open to the surface or just below the _____ on magnetic materials.

43. Radiographic inspection (RT) is a method for detecting flaws inside _____ _____.

44. Ultrasonic inspection (UT) is fast and uses few consumable supplies, which makes it _____ for schools to use.

45. Leak checking can be performed by filling the welded container with either a gas or _____.

46. Eddy current inspection (ET) is another nondestructive test; this method is based upon magnetic induction in which a magnetic field induces _____ currents within the material being tested.

47. Hardness is the resistance of metal to _____ and is an index of the wear resistance and strength of the metal.

FIGURE IDENTIFICATION EXERCISES

Figure 1-14

Instructions: In the space provided, identify the items shown in the illustration.

Name: _____ Date: _____

Class: _____ Instructor: _____ Score: _____

A _____

B _____

C _____

D _____

E _____

F _____

G _____

H _____

I _____

J _____

K _____

L _____

M _____

N _____

O _____

P _____

Q _____

R _____

Figure 1-14

Instructions: In the space provided, identify the items shown in the illustration.

Figure 1-15

Instructions: In the space provided, identify the items shown in the illustration.

Name: _____ Date: _____

Class: _____ Instructor: _____ Score: _____

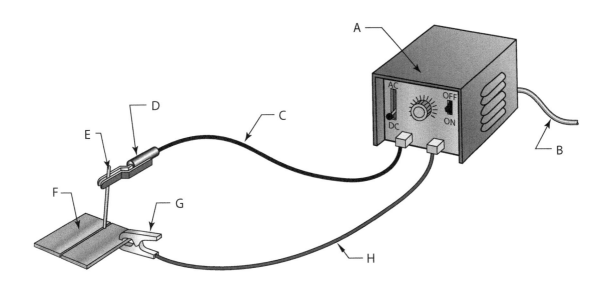

A _____

B _____

C _____

D _____

E _____

F _____

G _____

H _____

Figure 1-15

Instructions: In the space provided, identify the items shown in the illustration.

Name _____ Date _____

Class _____ Instructor _____ Score _____

A _____

B _____

C _____

D _____

E _____

F _____

G _____

H _____

Figure 1-16

Instructions: In the space provided, identify the items shown in the illustration.

Name:_____ Date:_____

Class:_____ Instructor:_____ Score:_____

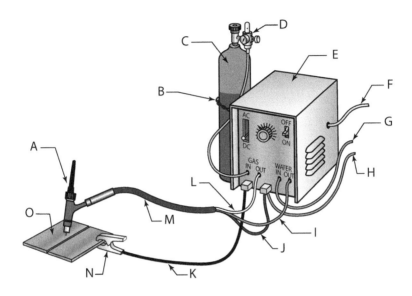

A _____

B _____

C _____

D _____

E _____

F _____

G _____

H _____

I _____

J _____

K _____

L _____

M_____

N _____

O _____

Figure 1-18

Instructions: In the space provided, identify the items shown in the illustration.

Name: _____ Date: _____

Class: _____ Instructor: _____ Score: _____

Figure 1-17

Instructions: In the space provided, identify the items shown in the illustration.

Name: _____ Date: _____

Class: _____ Instructor: _____ Score: _____

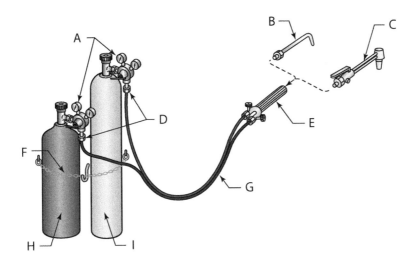

A _____

B _____

C _____

D _____

E _____

F _____

G _____

H _____

I _____

Figure 1-17

Instructions: In the space provided, identify the items shown in the illustration.

Name: _____ Date: _____

Class: _____ Instructor: _____ Score: _____

A. _____

B. _____

C. _____

D. _____

E. _____

F. _____

G. _____

H. _____

Figure 1-18

Instructions: In the space provided, identify the items shown in the illustration.

Name: _____ Date: _____

Class: _____ Instructor: _____ Score: _____

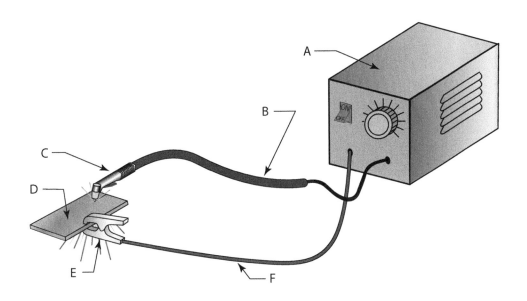

A _____

B _____

C _____

D _____

E _____

F _____

Figure 1-18

Figure Identification Exercises

Table 1-1A

Instructions: Identify the welding processes in the blanks to the right of each of the acronyms (abbreviations).

Name:_____ Date:_____

Class:_____ Instructor:_____ Score:_____

1. AHW _____
2. BMAW _____
3. CAW _____
4. CAW-G _____
5. CAW-S _____
6. CAW-T _____
7. EGW _____
8. FCAW _____

1. CEW _____
2. CW _____
3. DFW _____
4. EXW _____
5. FOW _____
6. FRW _____
7. HPW _____
8. ROW _____
9. USW _____

1. DS _____
2. FS _____
3. IS _____
4. IRS _____
5. INS _____
6. RS _____
7. TS _____
8. WS _____

1. FW _____
2. HFRW _____
3. PEW _____
4. PRW _____
5. RSEW _____
6. RSW _____
7. UW _____

1. EASP _____
2. FLSP _____
3. PSP _____

1. FOC _____
2. POC _____
3. OFC _____
4. OFC-A _____
5. OFC-H _____
6. OFC-N _____
7. OFC-P _____
8. AOC _____
9. OLC _____

ARC WELDING (AW)

SOLID STATE WELDING (SSW)

SOLDERING (S)

WELDING PROCESSES

RESISTANCE WELDING (RW)

THERMAL SPRAYING (THSP)

ALLIED PROCESSES

OXYGEN CUTTING (OC)

THERMAL CUTTING (TC)

OTHER CUTTING

Table 1-1B

Instructions: List the welding processes in the blanks to the right of each of the acronyms (abbreviations).

Name: _____ Date: _____

Class: _____ Instructor: _____ Score: _____

ARC WELDING (AW)

BRAZING (B)

WELDING PROCESSES

OTHER WELDING

OXYFUEL GAS WELDING (OFW)

ALLIED PROCESSES

ADHESIVE BONDING (ABD)

THERMAL CUTTING (TC)

ARC CUTTING (AC)

OTHER CUTTING

1. GMAW _____	6. PAW _____
2. GMAW-P _____	7. SMAW _____
3. GMAW-S _____	8. SW _____
	9. SAW _____
4. GTAW _____	10. SAW-S _____
5. GTAW-P _____	11. W-T _____

1. AB _____	6. FB _____
2. BB _____	7. IB _____
3. DFB _____	8. IRB _____
4. DB _____	9. TB _____
5. FLB _____	10. TCAB _____

1. EBW _____	5. ESW _____
2. EBW-HV _____	6. FLOW _____
3. EBW-MV _____	7. IW _____
	8. LBW _____
4. EBW-NV _____	9. TW _____

| 1. AAW _____ | 3. OHW _____ |
| 2. OAW _____ | 4. PGW _____ |

1. AAC _____	5. MAC _____
2. CAC _____	6. PAC _____
3. GMAC _____	7. SMAC _____
4. GTAC _____	

| 1. EBC _____ | 2. LBC _____ |

© Delmar Cengage Learning

Figures 2-1–2-3

Name:_____ Date:_____

Class:_____ Instructor:_____ Score:_____

A _____

B _____

C _____

D _____

E _____

F _____

G _____

H _____

I _____

J _____

K _____

L _____

M_____

N _____

O _____

P _____

Q _____

R _____

S _____

T _____

U _____

V _____

Figures 2-1-2-3

Instructions: In the space provided, identify the items shown in the illustrations.

Name _____ Date _____

Class _____ Instructor _____ Score _____

Figures 2-5, 2-6, and 2-8

Instructions: In the space provided, identify the items shown in the illustrations.

Name: _____ Date: _____

Class: _____ Instructor: _____ Score: _____

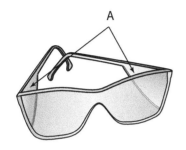

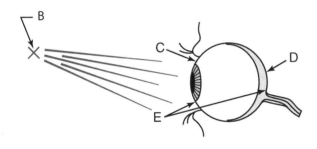

A _____

B _____

C _____

D _____

E _____

F _____

G _____

H _____

I _____

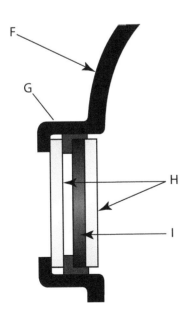

Figure 2-18

Instructions: In the space provided, identify the items shown in the illustration.

Name:_____ Date:_____

Class:_____ Instructor:_____ Score:_____

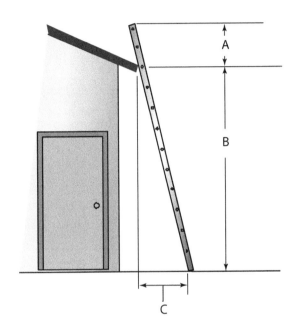

A _____

B _____

C _____

Figure 2-18

Instructions: In the space provided, identify the items shown in the illustration.

Name: _____ Date: _____

Class: _____ Instructor: _____ Score: _____

A. _____

B. _____

C. _____

Figures 2-29–2-31

Instructions: In the space provided, identify the items shown in the illustrations.

Name: _____ Date: _____

Class: _____ Instructor: _____ Score: _____

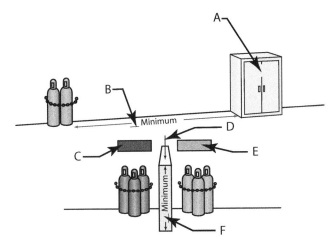

A _____

B _____

C _____

D _____

E _____

F _____

G _____

H _____

I _____

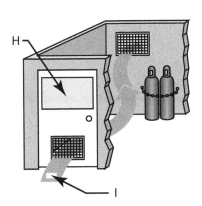

Figures 2-29—2-31

Instructions: In the space provided, identify the items shown in the illustrations.

Name: _____ Date: _____

Class: _____ Instructor: _____

Figures 2-34–2-37

Instructions: In the space provided, identify the items shown in the illustrations.

Name:_____ Date:_____

Class:_____ Instructor:_____ Score:_____

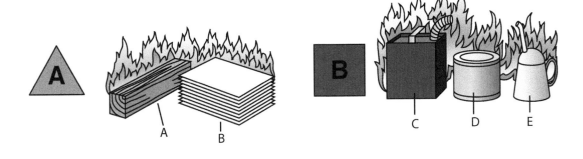

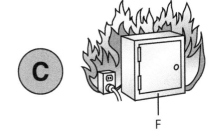

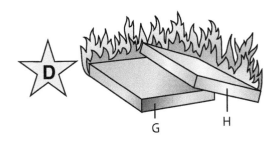

A _____

B _____

C _____

D _____

E _____

F _____

G _____

H _____

Figures 2-34–2-37

Instructions: In the space provided, identify the items shown in the illustrations.

Name: _____ Date: _____

Class: _____ Instructor: _____ Score: _____

A. _____

B. _____

C. _____

D. _____

E. _____

F. _____

G. _____

H. _____

Figure 2-38

Instructions: In the space provided, identify the items shown in the illustration.

Name: _____ Date: _____

Class: _____ Instructor: _____ Score: _____

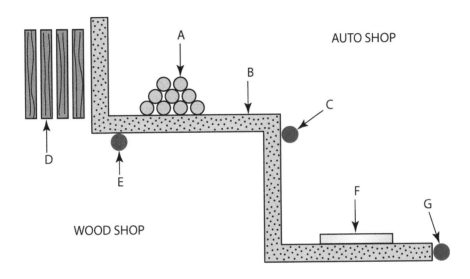

A _____

B _____

C _____

D _____

E _____

F _____

G _____

Figure 2-38

Instructions: In the space provided, identify the items shown in the illustration.

Name _____ Date _____

Date _____ Instructor _____ Score _____

Figure 2-52

Instructions: In the space provided, identify the items shown in the illustrations.

Name: _____ Date: _____

Class: _____ Instructor: _____ Score: _____

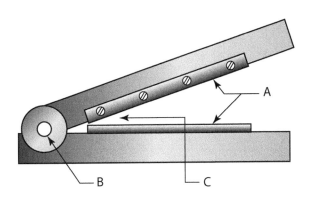

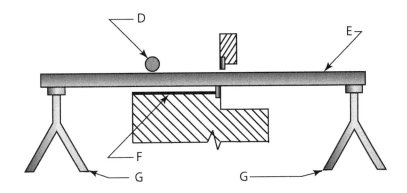

A _____

B _____

C _____

D _____

E _____

F _____

G _____

A. _____

B. _____

C. _____

D. _____

E. _____

F. _____

Figure 4-6

Instructions: In the space provided, identify the items shown in the illustrations.

Name: _____ Date: _____

Class: _____ Instructor: _____ Score: _____

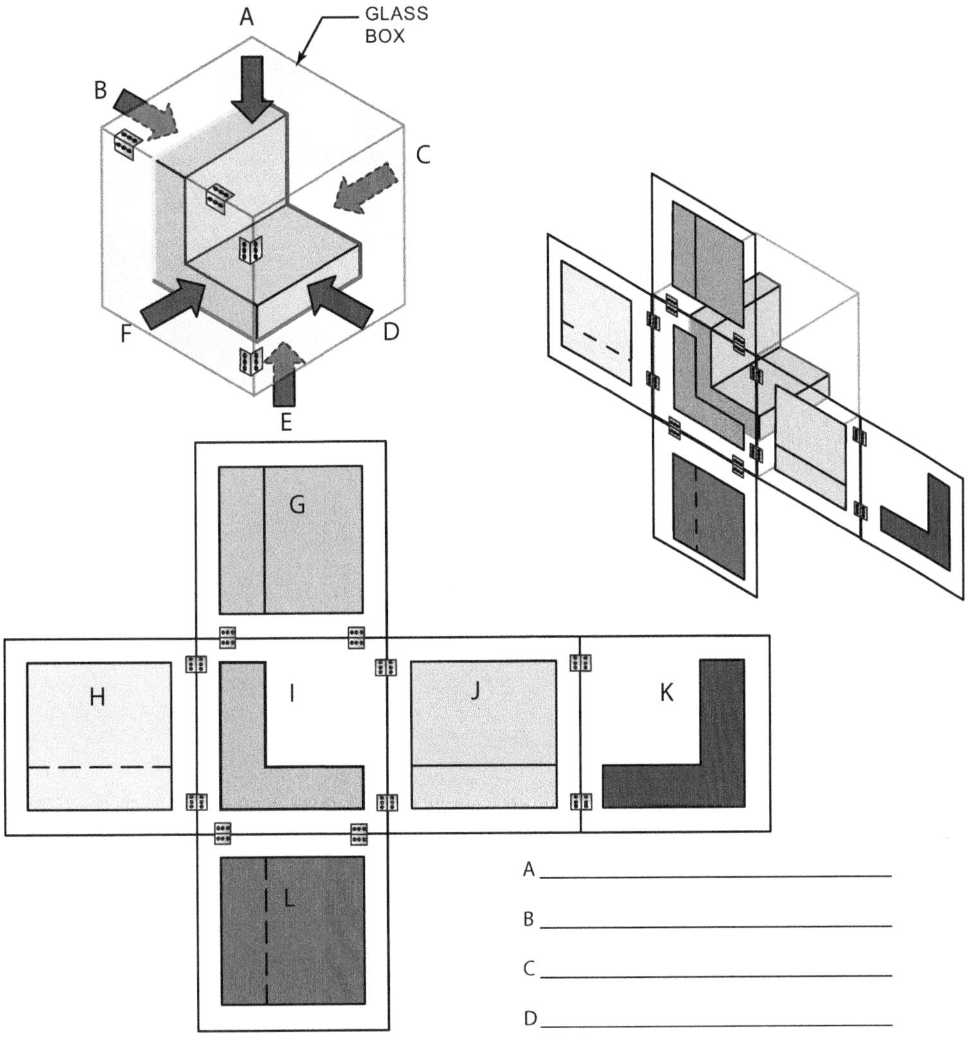

GLASS BOX

A _____

B _____

C _____

D _____

E _____

F _____

G _____

H _____

I _____

J _____

K _____

L _____

Figure 4–6

Instructions: In the space provided, identify the items shown in the illustration.

Name _____ Date _____

Class _____ Instructor _____ Phone _____

Figure 4-20

Instructions: In the space provided, identify the items shown in the illustrations.

Name: _____ Date: _____

Class: _____ Instructor: _____ Score: _____

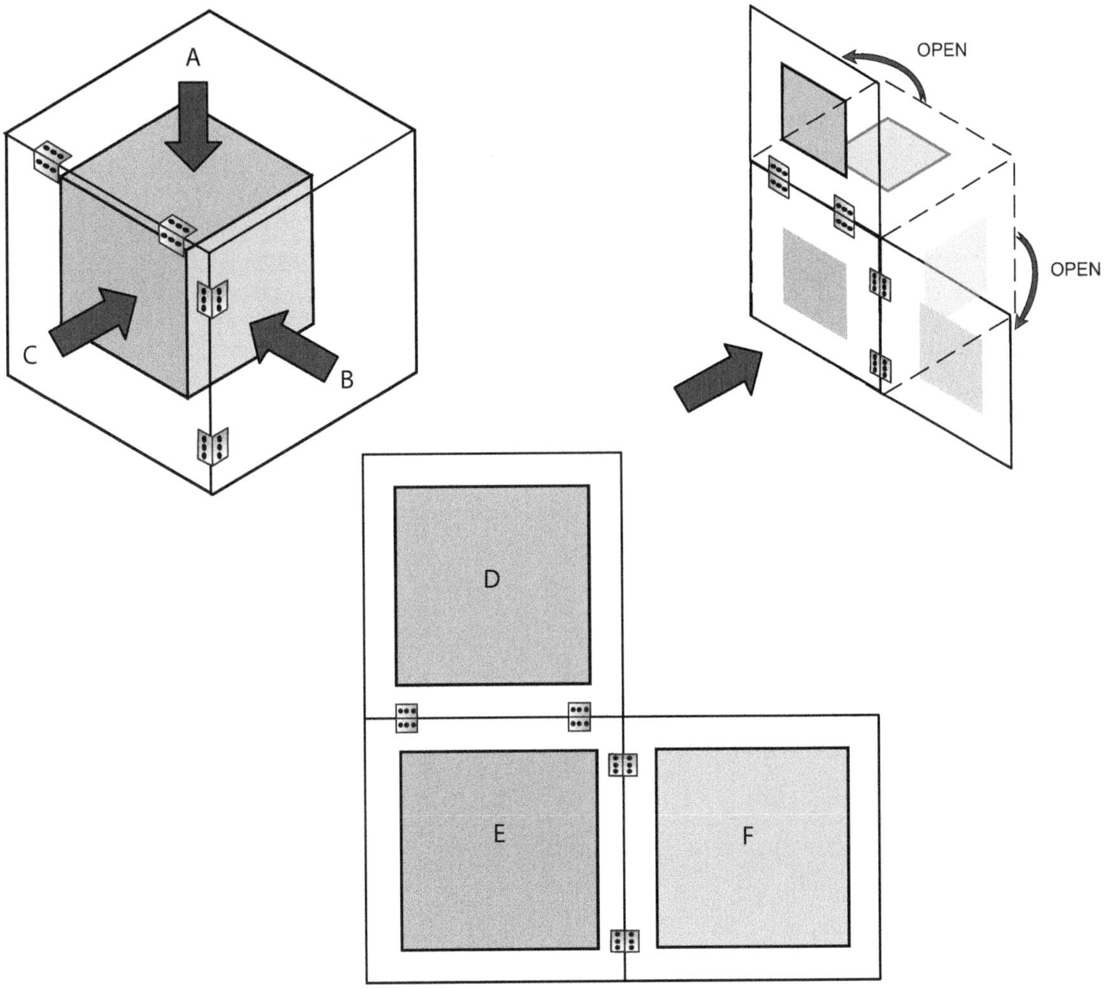

A _____

B _____

C _____

D _____

E _____

F _____

Figure 4-20

Instructions: In the space provided, identify the functions shown in the illustrations.

Name: _____ Date: _____

Class: _____ Instructor: _____ Score: _____

A. _____

B. _____

C. _____

D. _____

E. _____

F. _____

Figure 4-22

Instructions: In the space provided, identify the items shown in the illustrations.

Name: _____ Date: _____

Class: _____ Instructor: _____ Score: _____

A _____

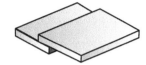

B _____

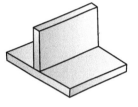

C _____

D _____

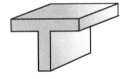

E _____

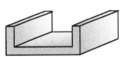

F _____

G _____

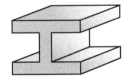

H _____

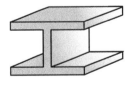

I _____

Figure 4-T1

Instructions: In the space provided, identify the items shown in the illustrations.

Name: _____ Date: _____

Class: _____ Instructor: _____ Score: _____

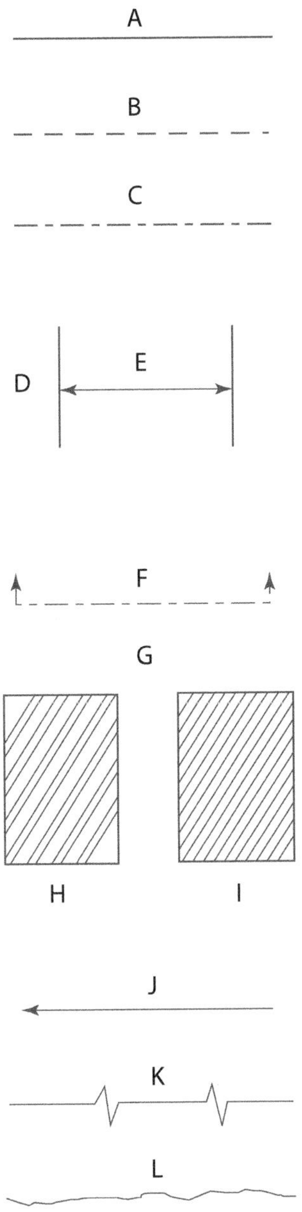

A _____

B _____

C - - - - - - - -

D - · - · - · - · -

A _____

B _____

C _____

D _____

E _____

F _____

G _____

H _____

I _____

J _____

K _____

L _____

D |← E →|

F

G

H   I

J ←_____

K

L

Figure 4-T1

Instructions: In the space provided, identify the items shown in the illustrations.

Name: _____ Date: _____

Class: _____ Instructor: _____ Score: _____

A _____

B _____

C _____

D _____

E _____

K _____

L _____

A _____

B _____

C _____

Figure 5-5

Instructions: In the space provided, identify the items shown in the illustrations.

Name: _____ Date: _____

Class: _____ Instructor: _____ Score: _____

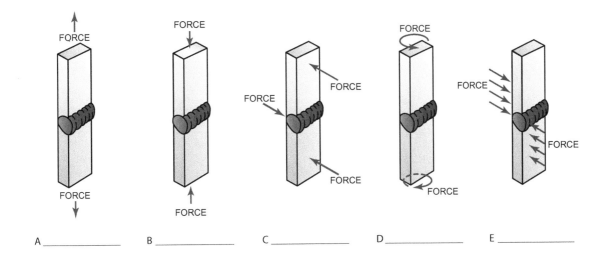

A _____   B _____   C _____   D _____   E _____

Instructions: In the space provided, identify the items shown in the illustration.

Name _____ Date _____

Class _____ Instructor _____ Score _____

Figure 5-6

Instructions: In the space provided, identify the items shown in the illustrations.

Name: _____ Date: _____

Class: _____ Instructor: _____ Score: _____

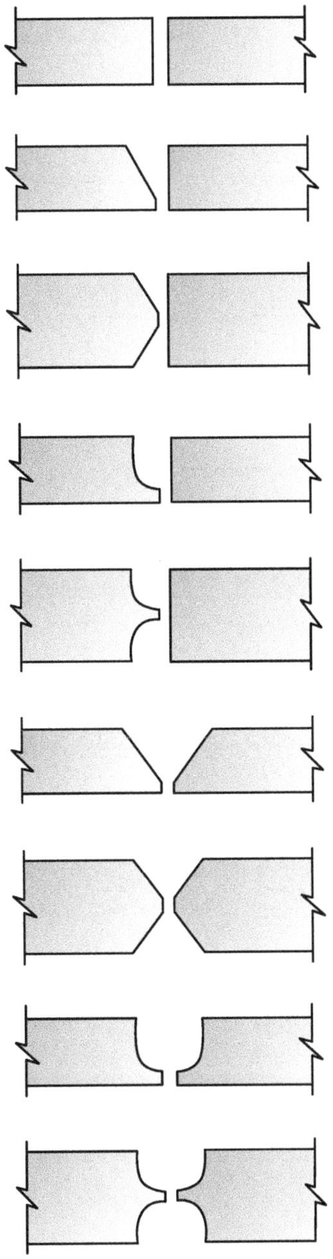

A _____

B _____

C _____

D _____

E _____

F _____

G _____

H _____

I _____

Figure 5-6

Instructions: In the space provided, identify the items shown in the illustration.

Name: _____ Class: _____

Class: _____ Instructor: _____ Score: _____

Figure 5-8

Instructions: In the space provided, identify the items shown in the illustrations.

Name: _____ Date: _____

Class: _____ Instructor: _____ Score: _____

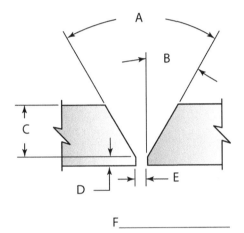

F_____

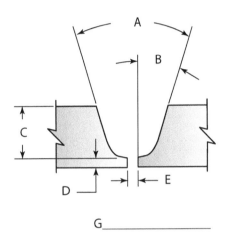

G_____

A _____

B _____

C _____

D _____

E _____

F _____

G _____

Figure 5-8

Instructions: In the space provided, identify the items shown in the illustration.

Name: _____ Date: _____

Class: _____ Instructor: _____ Score: _____

A. _____

B. _____

C. _____

D. _____

E. _____

F. _____

G. _____

Figure 5-12

Instructions: In the space provided, identify the items shown in the illustrations.

Name:_____ Date:_____

Class:_____ Instructor:_____ Score:_____

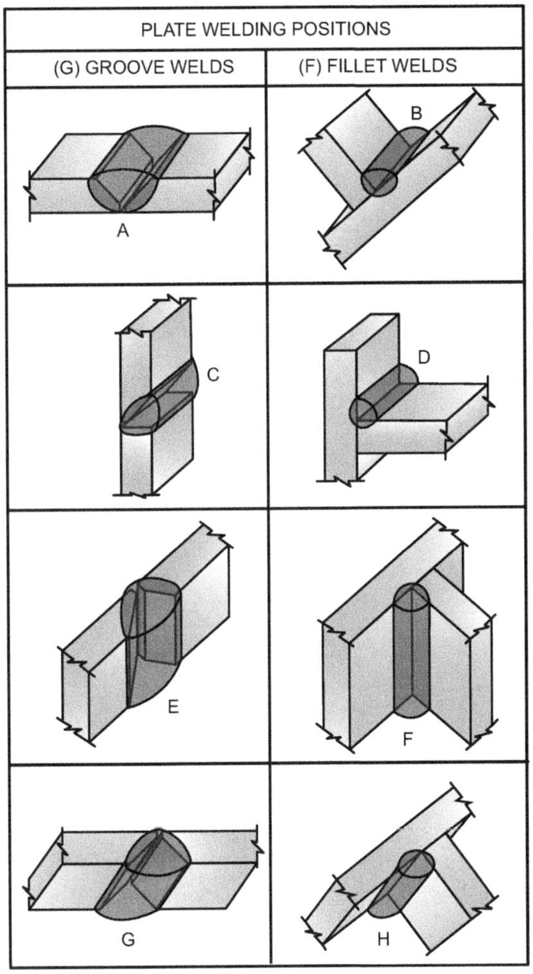

PLATE WELDING POSITIONS	
(G) GROOVE WELDS	(F) FILLET WELDS
A	B
C	D
E	F
G	H

A _____

B _____

C _____

D _____

E _____

F _____

G _____

H _____

Figure 5-17

Instructions: In the space provided, identify the items shown in the illustrations.

Name: _____ Date: _____

Class: _____ Instructor: _____ Score: _____

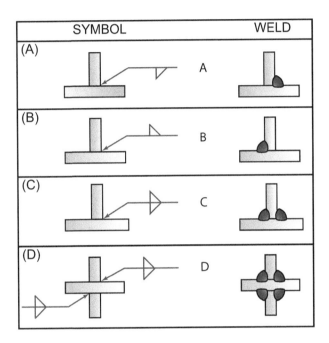

A _____

B _____

C _____

D _____

Figure 5-17

Instructions: In the space provided, identify the items shown in the illustration.

Name _____ Date _____

Class _____ Instructor _____ Score _____

SYMBOL	WELD
(A)	
(B)	
(C)	
(D)	

Figure 6-8A

Instructions: In the space provided, identify the items shown in the illustration.

Name: _____ Date: _____

Class: _____ Instructor: _____ Score: _____

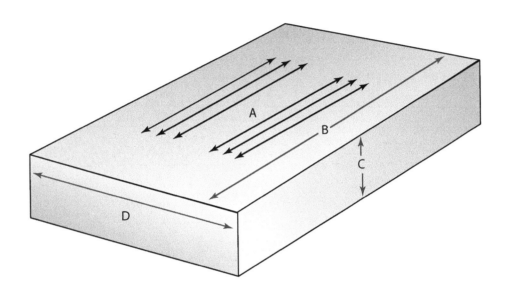

A _____

B _____

C _____

D _____

Figure 6-8A

Instructions: In the space provided, identify the items shown in the illustration.

Name _____ Date _____

Class _____ Instructor _____ Score _____

A _____

B _____

C _____

D _____

Figure 6-12

Instructions: In the space provided, identify the items shown in the illustration.

Name: _____ Date: _____

Class: _____ Instructor: _____ Score: _____

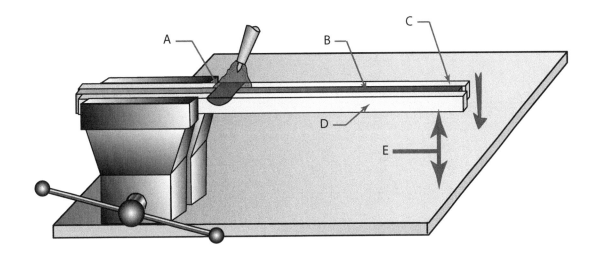

A _____

B _____

C _____

D _____

E _____

Figure 6-12

Instructions: In the space provided, identify the items shown in the illustration.

Name _____ Date _____

Class _____ Instructor _____ Score _____

A _____

B _____

C _____

D _____

Figure 6-14

Instructions: In the space provided, identify the items shown in the illustrations.

Name:_____ Date:_____

Class:_____ Instructor:_____ Score:_____

STEP 1
A

STEP 2
B

STEP 3
C

STEP 4
D

STEP 5
E

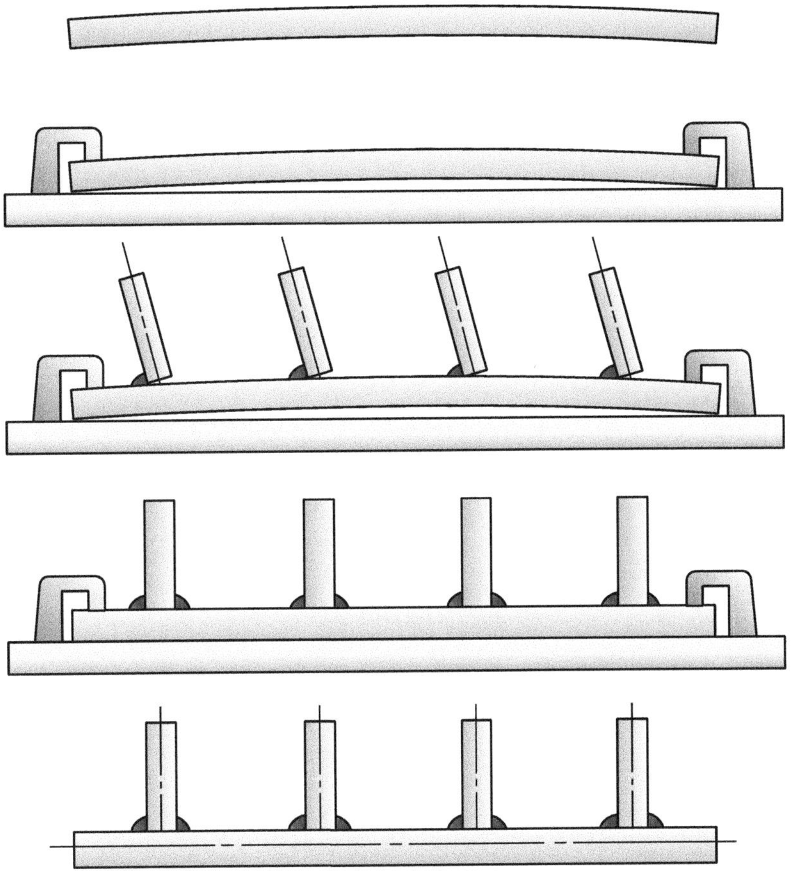

A _____

B _____

C _____

D _____

E _____

Figure 6-14

Instructions: In the space provided, identify the items shown in the illustrations.

Name: _____ Date: _____

Class: _____ Instructor: _____ Score: _____

STEP 1
A

STEP 2
B

STEP 3
C

STEP 4
D

STEP 5
E

A _____

B _____

C _____

D _____

E _____

Figure 6-24B

Instructions: In the space provided, identify the items shown in the illustrations.

Name: _____ Date: _____

Class: _____ Instructor: _____ Score: _____

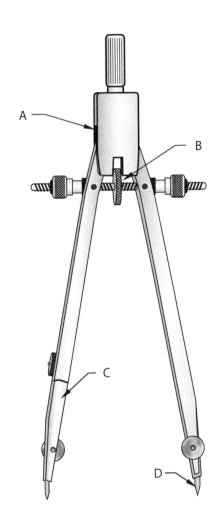

A _____

B _____

C _____

D _____

Figure 6-24B

Instructions: In the space provided, identify the items shown in the illustration.

Name _____ Date _____

Class _____ Instructor _____ Score _____

Figure 6-34

Instructions: In the space provided, identify the items shown in the illustration.

Name: _____ Date: _____

Class: _____ Instructor: _____ Score: _____

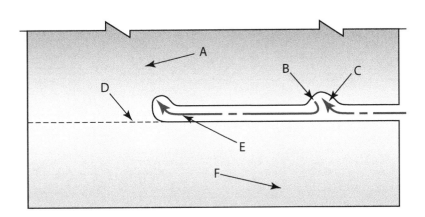

A _____

B _____

C _____

D _____

E _____

F _____

Figure 6-34

Instructions: In the space provided, identify the items shown in the illustration.

Name: _____ Date: _____

Class: _____ Instructor: _____ Score: _____

Figure 7-3

Instructions: In the space provided, identify the items shown in the illustration.

Name: _____ Date: _____

Class: _____ Instructor: _____ Score: _____

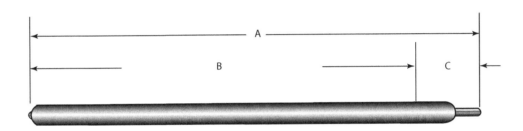

A _____

B _____

C _____

Figure 7-2

Instructions: In the space provided, identify the tissues shown in the illustration.

Name _____ Date _____

Class _____ Instructor _____ Score _____

Figure 7-5

Instructions: In the space provided, identify the items shown in the illustration.

Name:_____ Date:_____

Class:_____ Instructor:_____ Score:_____

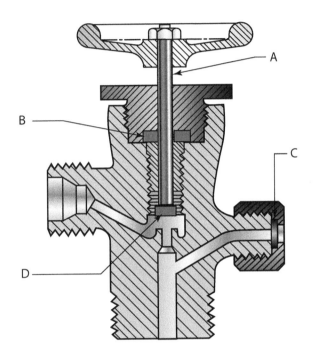

A _____

B _____

C _____

D _____

Figure 7-5

Instructions: In the space provided, identify the items shown in the illustration.

Name _____ Date _____

Class _____ Instructor _____ Score _____

A. _____

B. _____

C. _____

D. _____

Figure 7-6

Instructions: In the space provided, identify the items shown in the illustrations.

Name: _____ Date: _____

Class: _____ Instructor: _____ Score: _____

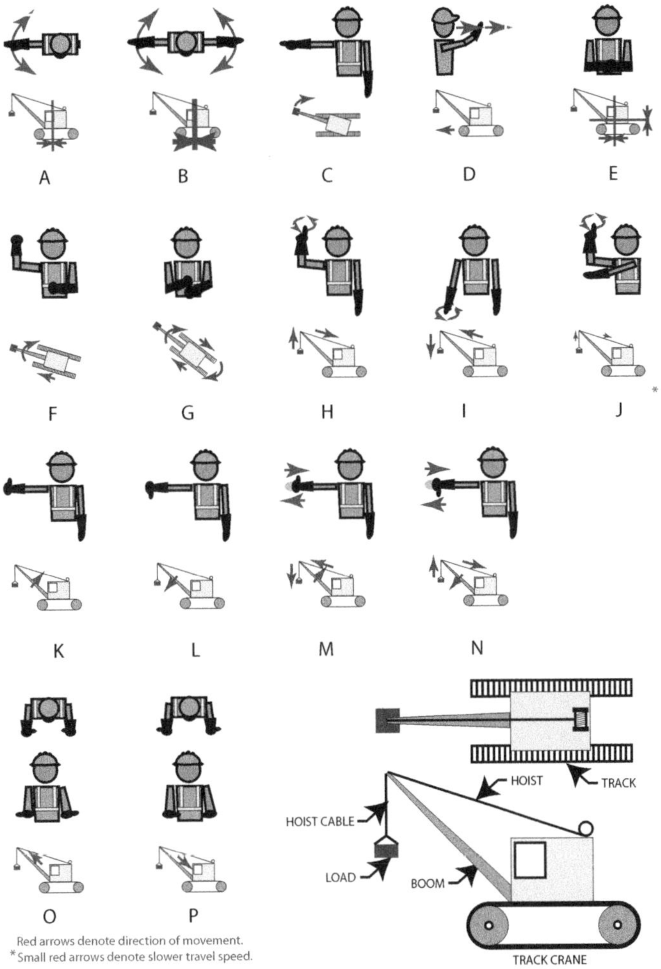

A   B   C   D   E

F   G   H   I   J

K   L   M   N

O   P

Red arrows denote direction of movement.
* Small red arrows denote slower travel speed.

HOIST

TRACK

HOIST CABLE

LOAD

BOOM

TRACK CRANE

A _____

B _____

C _____

D _____

E _____

F _____

G _____

H _____

I _____

J _____

K _____

L _____

M _____

N _____

O _____

P _____

Figure Identification Exercises

Figure 7-6

Instructions: In the space provided, identify the tools shown in the illustrations.

Name: _____ Date: _____

Class: _____ Instructor: _____ Score: _____

Figure 7-7

Name: _____ Date: _____

Class: _____ Instructor: _____ Score: _____

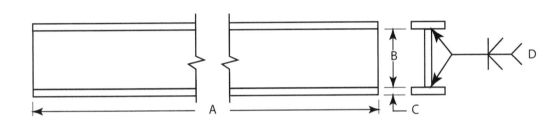

A _____

B _____

C _____

D _____

Figure 7-7

Instructions: In the space provided, identify the items shown in the illustration.

Name _____ Date _____

Class _____ Instructor _____ Score _____

A. _____

B. _____

C. _____

D. _____

Figure 8-1

Instructions: In the space provided, identify the items shown in the illustration.

Name: _____ Date: _____

Class: _____ Instructor: _____ Score: _____

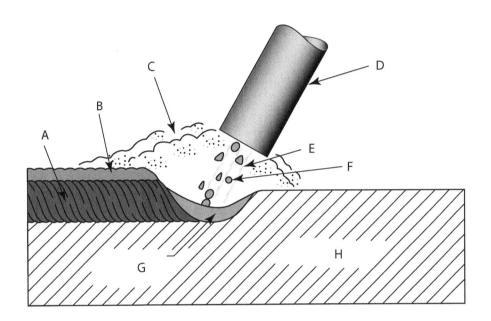

A _____

B _____

C _____

D _____

E _____

F _____

G _____

H _____

Figure 8-1

Instructions: In the space provided, identify the items shown in the illustration.

Name _____ Date _____

Instructor _____ Score _____

Figure 8-2

Name:_____ Date:_____

Class:_____ Instructor:_____ Score:_____

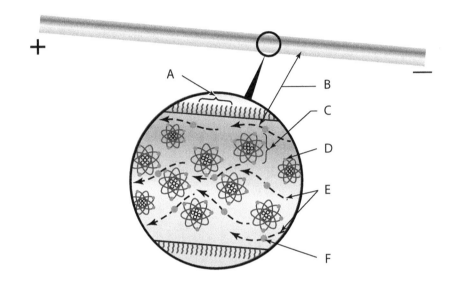

A_____

B_____

C_____

D_____

E_____

F_____

Figure 8-2

Instructions: In the space provided, identify the items shown in the illustration.

Name: _____ Date: _____

Class: _____ Instructor: _____ Score: _____

A. _____

B. _____

C. _____

D. _____

E. _____

Figure 8-3

Instructions: In the space provided, identify the items shown in the illustration.

Name: _____ Date: _____

Class: _____ Instructor: _____ Score: _____

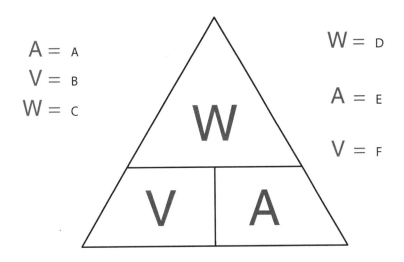

A = A

V = B

W = C

W = D

A = E

V = F

A _____

B _____

C _____

D _____

E _____

F _____

Figures 8-22, 8-24, and 8-25

Instructions: In the space provided, identify the items shown in the illustrations.

Name: _____ Date: _____

Class: _____ Instructor: _____ Score: _____

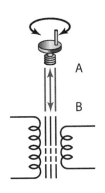

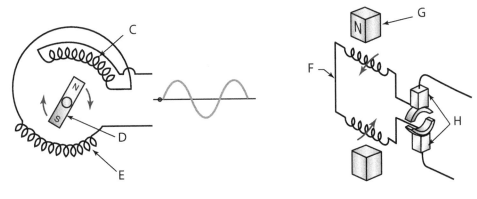

A _____

B _____

C _____

D _____

E _____

F _____

G _____

H _____

Figures 8-22, 8-24, and 8-25

Instructions: In the space provided, identify the items shown in the illustrations.

Name: _____ Date: _____

Date: _____ Instructor: _____ Score: _____

Figure 8-34

Instructions: In the space provided, identify the items shown in the illustration.

Name: _____ Date: _____

Class: _____ Instructor: _____ Score: _____

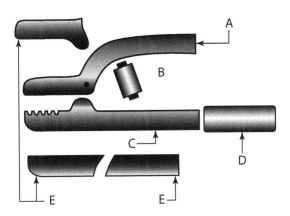

A _____

B _____

C _____

D _____

E _____

Figure 8-34

Instructions: In the space provided, identify the parts shown in the illustration.

Name _____ Date _____

Class _____ Instructor _____ Score _____

A. _____

B. _____

C. _____

D. _____

Figures 9-9 and 9-10

Instructions: In the space provided, identify the items shown in the illustrations.

Name:_____ Date:_____

Class:_____ Instructor:_____ Score:_____

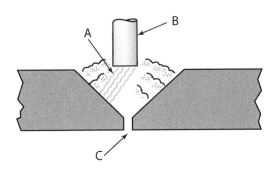

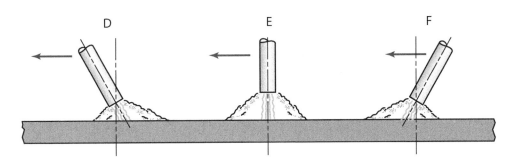

DIRECTION OF TRAVEL

A _____

B _____

C _____

D _____

E _____

F _____

Figures 9-9 and 9-10

Instructions: In the space provided, identify the items shown in the illustrations.

Name: _____ Date: _____

Instructor: _____ Score: _____

Class: _____

DIRECTION OF TRAVEL

A. _____

B. _____

C. _____

D. _____

E. _____

F. _____

Figures 9-11 and 9-12

Name: _____ Date: _____

Class: _____ Instructor: _____ Score: _____

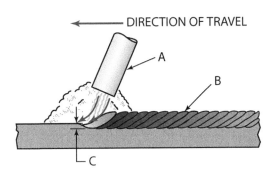

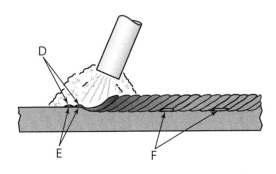

A _____

B _____

C _____

D _____

E _____

F _____

Figure 9-17

Instructions: In the space provided, identify the items shown in the illustrations.

Name: _____ Date: _____

Class: _____ Instructor: _____ Score: _____

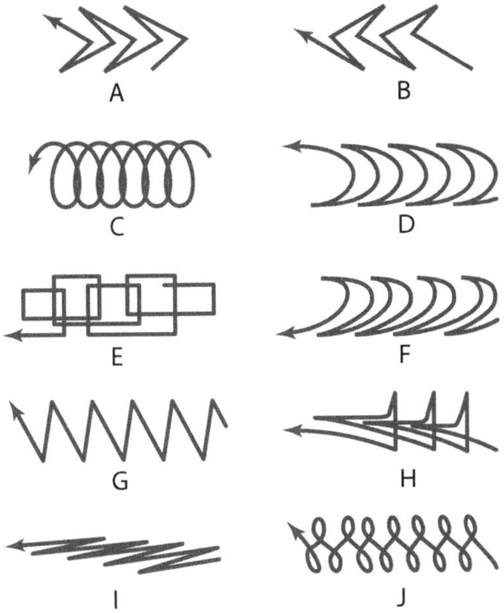

A _____

B _____

C _____

D _____

E _____

F _____

G _____

H _____

I _____

J _____

Figure 9-17

Instructions: In the space provided, identify the items shown in the illustrations.

Name: _____ Date: _____

Class: _____ Instructor: _____ Score: _____

Figure 10-4

Instructions: In the space provided, identify the items shown in the illustration.

Name: _____ Date: _____

Class: _____ Instructor: _____ Score: _____

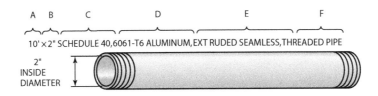

10' × 2" SCHEDULE 40, 6061-T6 ALUMINUM, EXT RUDED SEAMLESS, THREADED PIPE

2"
INSIDE
DIAMETER

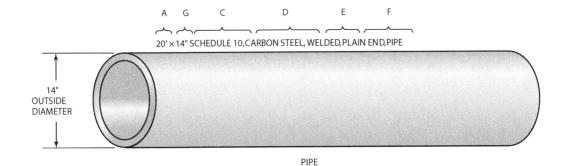

20' × 14" SCHEDULE 10, CARBON STEEL, WELDED, PLAIN END, PIPE

14"
OUTSIDE
DIAMETER

PIPE

A _____

B _____

C _____

D _____

E _____

F _____

G _____

Figure 10-4

Instructions: In the space provided, identify the items shown in the following illustration.

Name _____ Date _____

Class _____ Instructor _____ Score _____

Figure 10-5

Name: _____ Date: _____

Class: _____ Instructor: _____ Score: _____

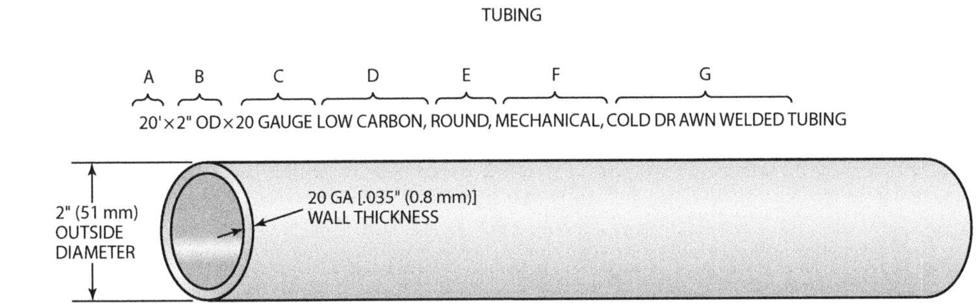

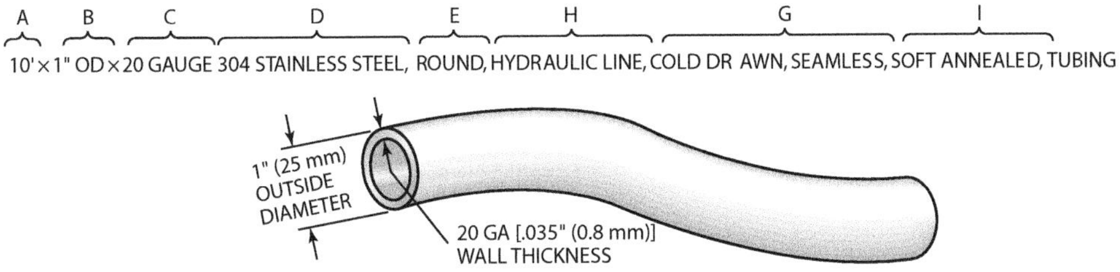

TUBING

A   B   C   D   E   F   G

20' × 2" OD × 20 GAUGE LOW CARBON, ROUND, MECHANICAL, COLD DRAWN WELDED TUBING

2" (51 mm)
OUTSIDE
DIAMETER

20 GA [.035" (0.8 mm)]
WALL THICKNESS

A   B   C   D   E   H   G   I

10' × 1" OD × 20 GAUGE 304 STAINLESS STEEL, ROUND, HYDRAULIC LINE, COLD DRAWN, SEAMLESS, SOFT ANNEALED, TUBING

1" (25 mm)
OUTSIDE
DIAMETER

20 GA [.035" (0.8 mm)]
WALL THICKNESS

A _____

B _____

C _____

D _____

E _____

F _____

G _____

H _____

I _____

Figure 10-5

Instructions: In the space provided, identify the items shown in the illustration.

Name: _____ Date: _____

Class: _____ Score: _____ Instructor: _____

TUBING

A ____ B ____ C ____ D ____ E ____ F ____

Figure 10-9

Instructions: In the space provided, identify the items shown in the illustration.

Name:______________________________ Date:______________

Class:______________ Instructor:______________________________ Score:______________

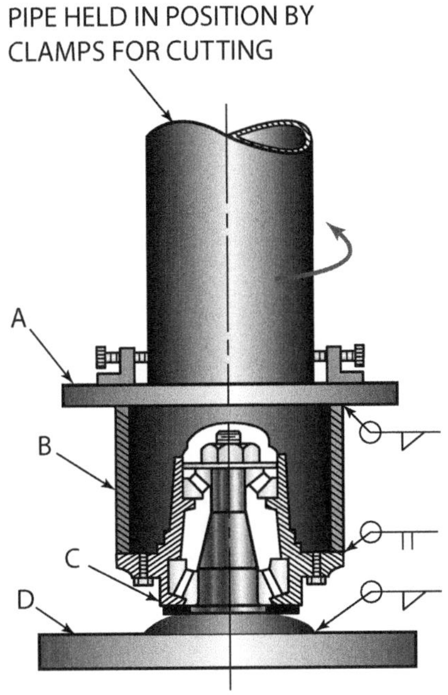
PIPE HELD IN POSITION BY
CLAMPS FOR CUTTING
A
B
C
D

A ______________________________

B ______________________________

C ______________________________

D ______________________________

Figure 10-13

Instructions: In the space provided, identify the items shown in the illustration.

Name:_____ Date:_____

Class:_____ Instructor:_____ Score:_____

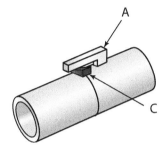

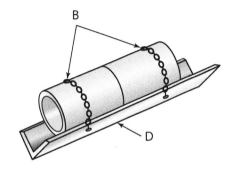

A _____

B _____

C _____

D _____

Figure 10-13

Instructions: In the space provided, identify the items shown in the illustration.

Name: _____ Date: _____

Class: _____ Instructor: _____ Score: _____

A. _____

B. _____

C. _____

D. _____

Figure 10-39

Instructions: In the space provided, identify the items shown in the illustration.

Name: _____ Date: _____

Class: _____ Instructor: _____ Score: _____

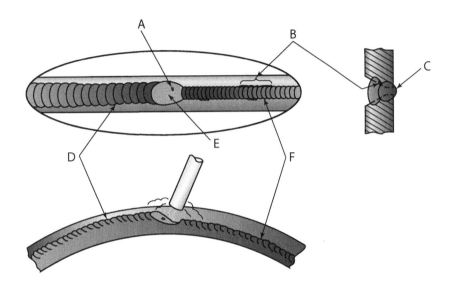

A _____

B _____

C _____

D _____

E _____

F _____

Figure 10-35

Instructions: In the space provided, identify the items shown in the illustration.

Name _____ Date _____

Class _____ Instructor _____ Score _____

Figure 11-1

Instructions: In the space provided, identify the items shown in the illustration.

Name: _____ Date: _____

Class: _____ Instructor: _____ Score: _____

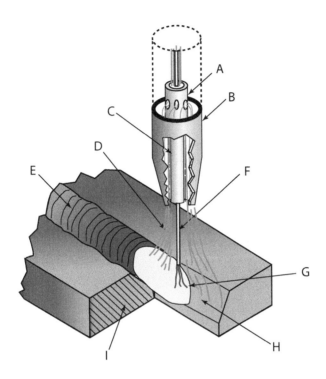

A _____

B _____

C _____

D _____

E _____

F _____

G _____

H _____

I _____

Figure 11-1

Instructions: In the space provided, identify the items shown in the illustration.

Name: _____ Date: _____

Class: _____ Instructor: _____ Score: _____

A. _____

B. _____

C. _____

D. _____

E. _____

F. _____

G. _____

H. _____

I. _____

Figure 11-3

Instructions: In the space provided, identify the items shown in the illustration.

Name:_____ Date:_____

Class:_____ Instructor:_____ Score:_____

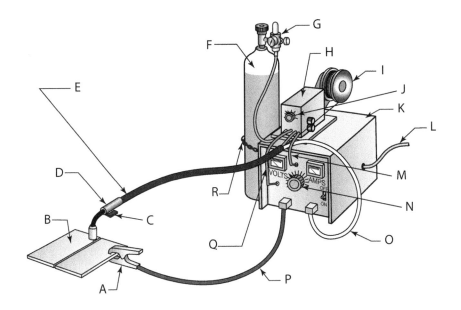

A_____   J_____

B_____   K_____

C_____   L_____

D_____   M_____

E_____   N_____

F_____   O_____

G_____   P_____

H_____   Q_____

I_____   R_____

Figure 11-3

Instructions: In the space provided, identify the items shown in the illustration.

Name _____ Date _____

Class _____ Instructor _____ Score _____

A. _____

B. _____

C. _____

D. _____

E. _____

F. _____

G. _____

H. _____

I. _____

J. _____

K. _____

L. _____

M. _____

N. _____

O. _____

Figure 11-6

Instructions: In the space provided, identify the items shown in the illustration.

Name: _____ Date: _____

Class: _____ Instructor: _____ Score: _____

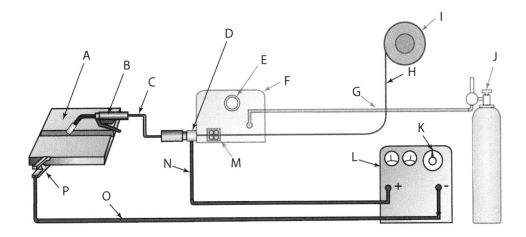

A _____   I _____

B _____   J _____

C _____   K _____

D _____   L _____

E _____   M _____

F _____   N _____

G _____   O _____

H _____   P _____

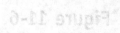

Figure 11-6

Instructions: In the space provided, identify the items shown in the illustration.

Name: _____ Date: _____

Class: _____ Instructor: _____ Score: _____

Figures 11-8, 11-10, and 11-14

Instructions: In the space provided, identify the items shown in the illustrations.

Name: _____ Date: _____

Class: _____ Instructor: _____ Score: _____

A   B   C

D

E

F   G   H

I

J   K   L

M   N

O

P

A _____

B _____

C _____

D _____

E _____

F _____

G _____

H _____

I _____

J _____

K _____

L _____

M _____

N _____

O _____

P _____

Figures 11-15, 11-17, and 11-18

Instructions: In the space provided, identify the items shown in the illustrations.

Name:_____ Date:_____

Class:_____ Instructor:_____ Score:_____

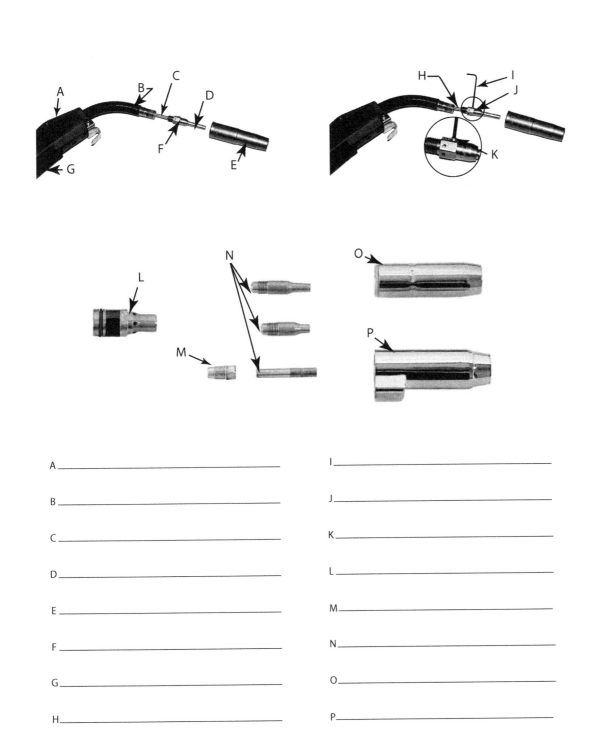

A_____ I_____

B_____ J_____

C_____ K_____

D_____ L_____

E_____ M_____

F_____ N_____

G_____ O_____

H_____ P_____

Figure Identification Exercises

Figures 11-15, 11-17, and 11-18

Instructions: In the space provided, identify the items shown in the illustrations.

Name: _____ Date _____

Class _____ Instructor _____ Score _____

Figure 11-32

Name: _____ Date: _____

Class: _____ Instructor: _____ Score: _____

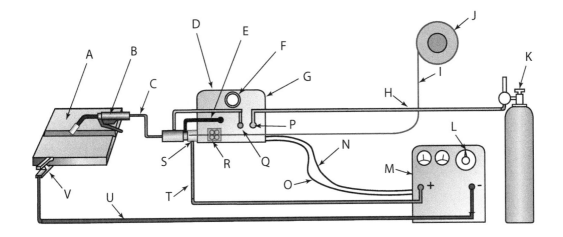

A _____   L _____

B _____   M _____

C _____   N _____

D _____   O _____

E _____   P _____

F _____   Q _____

G _____   R _____

H _____   S _____

I _____   T _____

J _____   U _____

K _____   V _____

Figure 11-32

Instructions: In the space provided, identify the items shown in the illustration.

Name: _____ Date: _____

Class: _____ Instructor: _____ Score: _____

Figures 12-1 and 12-2

Instructions: In the space provided, identify the items shown in the illustrations.

Name:_____ Date:_____

Class:_____ Instructor:_____ Score:_____

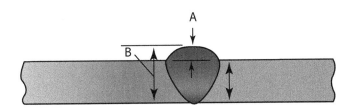

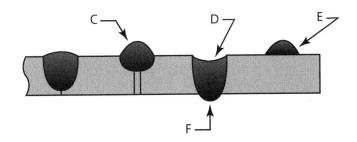

A _____

B _____

C _____

D _____

E _____

F _____

Figures 12-1 and 12-2

Instructions: In the space provided, identify the items shown in the illustrations.

Name _____ Date _____

Class _____ Instructor _____ Score _____

Figures 12-3 and 12-6

Instructions: In the space provided, identify the items shown in the illustrations.

Name: _____ Date: _____

Class: _____ Instructor: _____ Score: _____

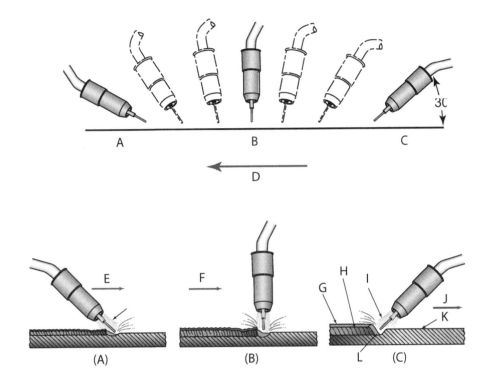

A _____

B _____

C _____

D _____

E _____

F _____

G _____

H _____

I _____

J _____

K _____

L _____

Figures 12-5 and 12-6

Instructions: In the space provided, identify the items shown in the illustrations.

Name: _____ Date: _____

Class: _____ Instructor: _____ Score: _____

A. _____

B. _____

C. _____

D. _____

E. _____

F. _____

G. _____

H. _____

I. _____

J. _____

K. _____

Figures 12-7 and 12-8

Instructions: In the space provided, identify the items shown in the illustrations.

Name:_____ Date:_____

Class:_____ Instructor:_____ Score:_____

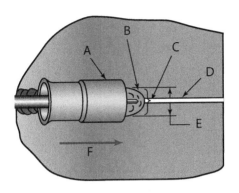

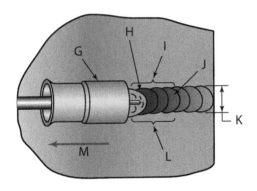

A _____

B _____

C _____

D _____

E _____

F _____

G _____

H _____

I _____

J _____

K _____

L _____

M_____

A. _____

B. _____

C. _____

D. _____

E. _____

F. _____

G. _____

H. _____

I. _____

J. _____

K. _____

L. _____

M. _____

Figures 12-9–12-11

Instructions: In the space provided, identify the items shown in the illustrations.

Name:_____ Date:_____

Class:_____ Instructor:_____ Score:_____

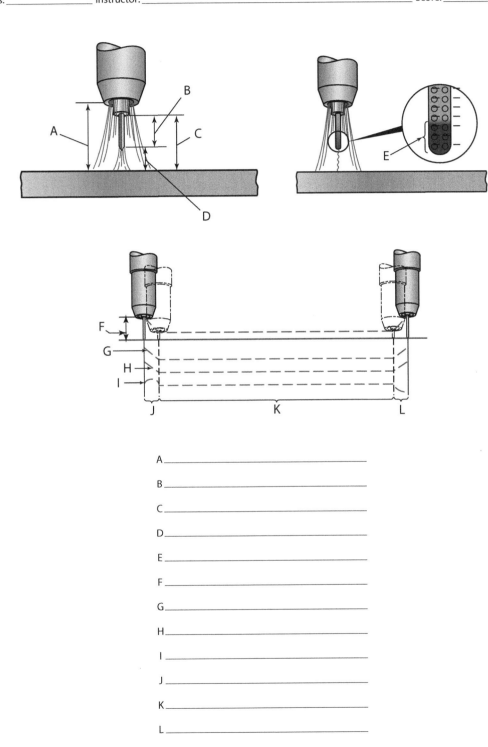

A_____

B_____

C_____

D_____

E_____

F_____

G_____

H_____

I_____

J_____

K_____

L_____

Figures 12-9–12-11

Instructions: In the space provided, identify the items shown in the illustrations.

Name: _____ Date: _____

Class: _____ Instructor: _____ Score: _____

Figures 12-22, 12-24, 12-26, and 12-27

Instructions: In the space provided, identify the items shown in the illustrations.

Name: _____ Date: _____

Class: _____ Instructor: _____ Score: _____

A _____

B _____

C _____

D _____

E _____

F _____

_____

G _____

_____

H _____

_____

Figures 12-22, 12-24, 12-26, and 12-27

Instructions: In the space provided, identify the items shown in the illustrations.

Name_____ Date_____

Class_____ Instructor_____ Score_____

A_____

B_____

C_____

D_____

E_____

F_____

G_____

H_____

Figures 12-36–12-39

Instructions: In the space provided, identify the items shown in the illustrations.

Name: _____ Date: _____

Class: _____ Instructor: _____ Score: _____

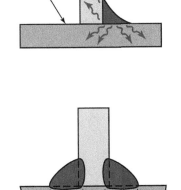

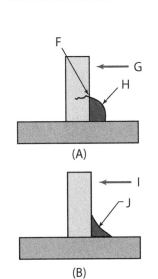

A _____

B _____

C _____

D _____

E _____

F _____

G _____

H _____

I _____

J _____

Figures 12-36—12-39

Instructions: In the space provided, identify the items shown in the illustrations.

Name: _____ Date: _____

Class: _____ Instructor: _____ Score: _____

Figures 12-41, 12-62, and 12-64

Instructions: In the space provided, identify the items shown in the illustrations.

Name: _____ Date: _____

Class: _____ Instructor: _____ Score: _____

A _____

B _____

C _____

D _____

E _____

F _____

G _____

H _____

I _____

J _____

K _____

L _____

M_____

Figures 12-41, 12-62, and 12-64

Instructions: In the space provided, identify the items shown in the illustrations.

Name _____ Date _____

Class _____ Instructor _____ Score _____

A. _____
B. _____
C. _____
D. _____
E. _____
F. _____
G. _____
H. _____
I. _____
J. _____
K. _____
L. _____
M. _____

Figure 13-1A

Instructions: In the space provided, identify the items shown in the illustration.

Name: _____ Date: _____

Class: _____ Instructor: _____ Score: _____

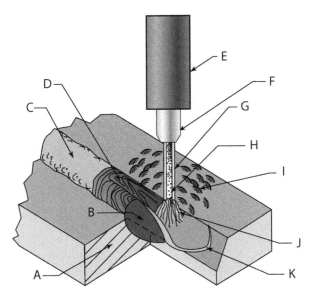

SELF-SHIELDED FLUX CORED ARC WELDING (FCAW-S)

A _____

B _____

C _____

D _____

E _____

F _____

G _____

H _____

I _____

J _____

K _____

Figure 13-1A

Instructions: In the space provided, identify the items shown in the illustration.

Name _____ Date _____

Class _____ Instructor _____ Score _____

Figure 13-1B

Instructions: In the space provided, identify the items shown in the illustration.

Name: _____ Date: _____

Class: _____ Instructor: _____ Score: _____

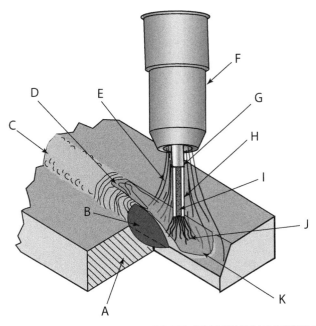

(A) DUAL-SHIELDED FLUX CORED ARC WELDING (FCAW-G)

A _____

B _____

C _____

D _____

E _____

F _____

G _____

H _____

I _____

J _____

K _____

Figure 13-18

Instructions: In the space provided, identify the items shown in the illustration.

Name _____ Date _____

Class _____ Instructor _____ Score _____

A. _____

B. _____

C. _____

D. _____

E. _____

F. _____

G. _____

H. _____

I. _____

J. _____

K. _____

Figures 13-3, 13-7, and 13-8

Instructions: In the space provided, identify the items shown in the illustrations.

Name: _____ Date: _____

Class: _____ Instructor: _____ Score: _____

A _____

B _____

C _____

D _____

E _____

F _____

G _____

H _____

I _____

J _____

K _____

L _____

M _____

Figures 13-9, 13-11, and 13-12

Instructions: In the space provided, identify the items shown in the illustrations.

Name:_____ Date:_____

Class:_____ Instructor:_____ Score:_____

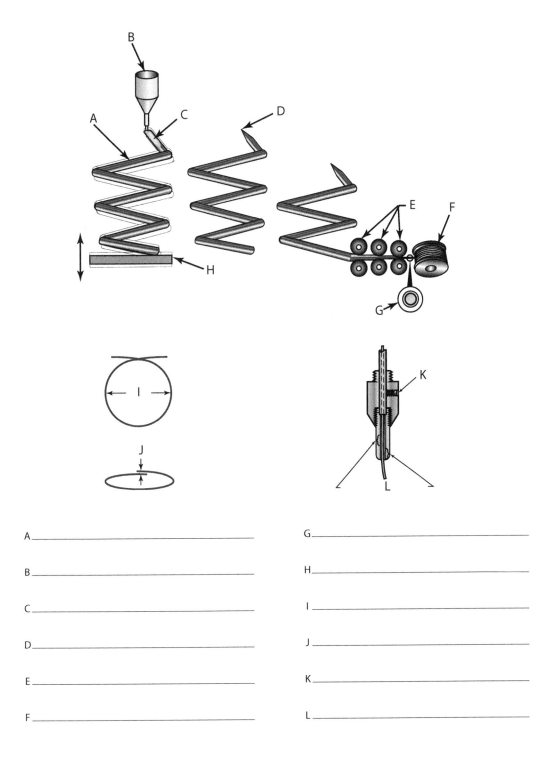

A_____   G_____

B_____   H_____

C_____   I_____

D_____   J_____

E_____   K_____

F_____   L_____

Figures 13-9, 13-11, and 13-12

Instructions: In the space provided, identify the items shown in the illustrations.

Name _____ Date _____

Class _____ Instructor _____

Figures 13-21 and 13-22

Instructions: In the space provided, identify the items shown in the illustrations.

Name: _____ Date: _____

Class: _____ Instructor: _____ Score: _____

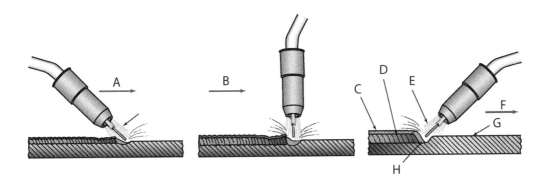

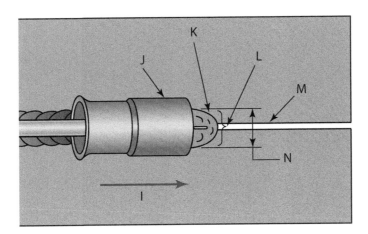

A_____

B_____

C_____

D_____

E_____

F_____

G_____

H_____

I_____

J_____

K_____

L_____

M_____

N_____

Figures 13-21 and 13-22

Instructions: In the space provided, identify the items shown in the illustrations.

Name _____ Date _____

Class _____ Instructor: _____ Time: _____

Figures 13-23 and 13-24

Instructions: In the space provided, identify the items shown in the illustrations.

Name:_____ Date:_____

Class:_____ Instructor:_____ Score:_____

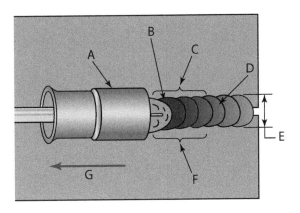

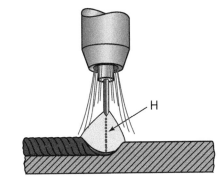

A_____

B_____

C_____

D_____

E_____

F_____

G_____

H_____

Figures 13-23 and 13-24

Instructions: In the space provided, identify the items shown in the illustrations.

Name: _____ Date: _____

Class: _____ Instructor: _____ Score: _____

A. _____

B. _____

C. _____

D. _____

E. _____

F. _____

G. _____

H. _____

Figures 13-25 and 13-26

Instructions: In the space provided, identify the items shown in the illustrations.

Name: _____ Date: _____

Class: _____ Instructor: _____ Score: _____

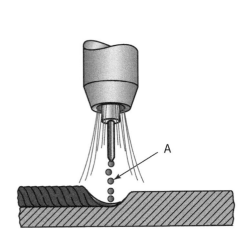

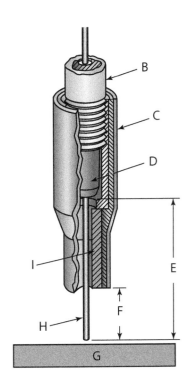

A _____

B _____

C _____

D _____

E _____

F _____

G _____

H _____

I _____

Figures 13-25 and 13-26

Instructions: In the space provided, identify the items shown in the illustrations.

Name _____ Date _____

Class _____ Instructor _____ Score _____

Figures 14-1 and 14-2

Instructions: In the space provided, identify the items shown in the illustrations.

Name:_____ Date:_____

Class:_____ Instructor:_____ Score:_____

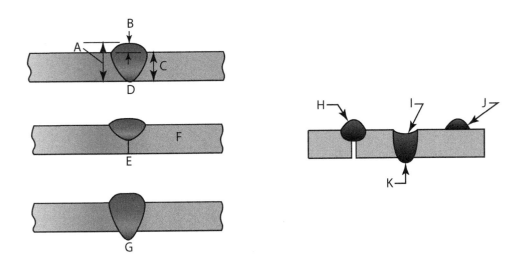

A_____

B_____

C_____

D_____

E_____

F_____

G_____

H_____

I_____

J_____

K_____

Figures 14-1 and 14-2

Instructions: In the space provided, identify the items shown in the illustrations.

Name _____ Date _____

Class _____ Instructor _____ Score _____

1. _____

2. _____

3. _____

4. _____

5. _____

6. _____

Figures 14-3, 14-4, 14-6, and 14-7

Instructions: In the space provided, identify the items shown in the illustrations.

Name: _____ Date: _____

Class: _____ Instructor: _____ Score: _____

A _____

B _____

C _____

D _____

E _____

F _____

G _____

H _____

I _____

J _____

K _____

Figures 14-3, 14-5, 14-6, and 14-7

Instructions: In the space provided identify the items shown in the illustrations.

Name: _____ Date: _____

Class: _____ Instructor: _____ Score: _____

Figures 14-9, 14-11, and 14-30

Instructions: In the space provided, identify the items shown in the illustrations.

Name: _____ Date: _____

Class: _____ Instructor: _____ Score: _____

A _____

B _____

C _____

D _____

E _____

F _____

G _____

H _____

I _____

J _____

K _____

L _____

Figures 14-9, 14-11, and 14-30

Instructions: In the space provided, identify the items shown in the illustrations.

Name: _____   Date: _____

Class: _____   Instructor: _____   Score: _____

A. _____

B. _____

C. _____

D. _____

E. _____

F. _____

G. _____

Figures 15-1 and 15-6

Instructions: In the space provided, identify the items shown in the illustrations.

Name: _____ Date: _____

Class: _____ Instructor: _____ Score: _____

A _____   H _____

B _____   I _____

C _____   J _____

D _____   K _____

E _____   L _____

F _____   M _____

G _____

Figures 15-1 and 15-6

Instructions: In the space provided, identify the items shown in the figures.

Name: _____ Date: _____

Class: _____ Instructor: _____ Score: _____

A. _____

B. _____

C. _____

D. _____

E. _____

Figure 15-7

Instructions: In the space provided, identify the items shown in the illustration.

Name: _____ Date: _____

Class: _____ Instructor: _____ Score: _____

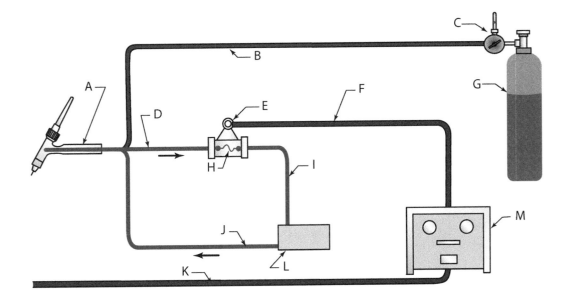

A _____   H _____

B _____   I _____

C _____   J _____

D _____   K _____

E _____   L _____

F _____   M _____

G _____

Figure 15-7

Instructions: In the space provided, identify the items shown in the illustration.

Name: _____ Date: _____

Class: _____ Score: _____ Instructor: _____

A. _____

B. _____

C. _____

D. _____

E. _____

M. _____

Figures 15-10–15-12

Instructions: In the space provided, identify the items shown in the illustrations.

Name: _____ Date: _____

Class: _____ Instructor: _____ Score: _____

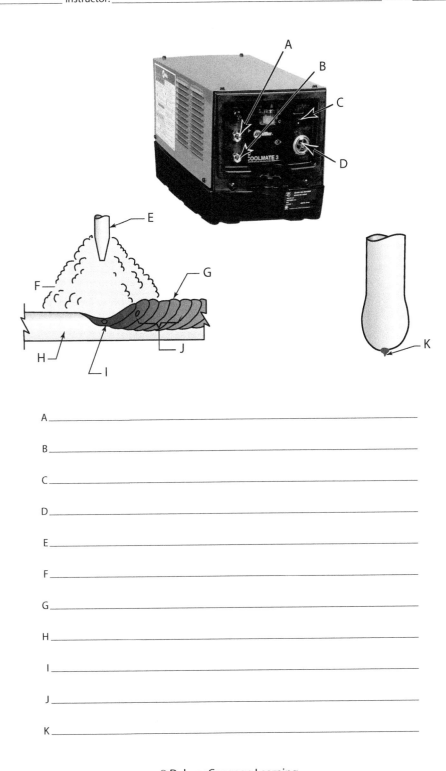

A _____

B _____

C _____

D _____

E _____

F _____

G _____

H _____

I _____

J _____

K _____

Figures 15-10-15-12

Instructions: In the space provided, identify the items shown in the illustrations.

Name _____ Date _____

Class _____ Instructor _____ Grade _____

Figures 15-18 and 15-20

Instructions: In the space provided, identify the items shown in the illustrations.

Name: _____ Date: _____

Class: _____ Instructor: _____ Score: _____

A_____	I_____
B_____	J_____
C_____	K_____
D_____	L_____
E_____	M_____
F_____	N_____
G_____	O_____
H_____	

Figures 15-18 and 15-20

Instructions: In the space provided, identify the items shown in the illustrations.

Name _____ Date _____

Class _____ Instructor _____

A. _____

B. _____

K. _____

L. _____

M. _____

N. _____

O. _____

Figures 15-21, 15-24, and 15-44

Instructions: In the space provided, identify the items shown in the illustrations.

Name: _____ Date: _____

Class: _____ Instructor: _____ Score: _____

Sine wave of alternating current at 60 cycles

A _____   G _____

B _____   H _____

C _____   I _____

D _____   J _____

E _____   K _____

F _____   L _____

Figures 16-1, 16-5, and 16-8

Instructions: In the space provided, identify the items shown in the illustrations.

Name:_____ Date:_____

Class:_____ Instructor:_____ Score:_____

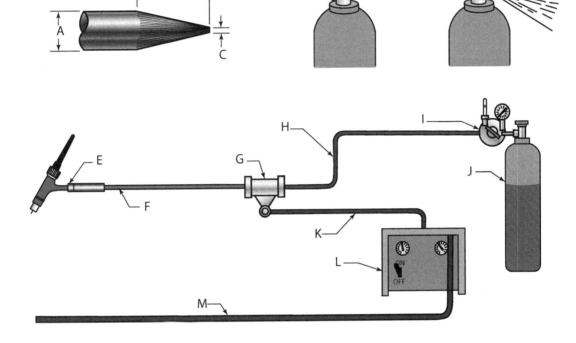

A _____

B _____

C _____

D _____

E _____

F _____

G _____

H _____

I _____

J _____

K _____

L _____

M_____

Figures 16-1, 16-5, and 16-8

Instructions: In the space provided, identify the items shown in the illustrations.

Name _____ Date _____

Class _____ Instructor _____ Score _____

A _____   H _____

B _____   I _____

J _____   K _____

D _____   L _____

E _____   M _____

_____   _____

Figures 16-17, 16-23, 16-37, and 16-40

Instructions: In the space provided, identify the items shown in the illustrations.

Name: _____ Date: _____

Class: _____ Instructor: _____ Score: _____

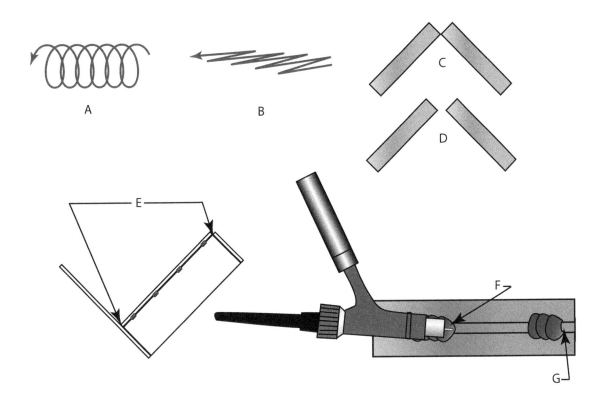

A _____

B _____

C _____

D _____

E _____

F _____

G _____

Figures 16-17, 16-23, 16-37, and 16-40

Instructions: In the space provided, identify the items shown in each illustration.

Figure 17-1

Instructions: In the space provided, identify the items shown in the illustration.

Name:_____ Date:_____

Class:_____ Instructor:_____ Score:_____

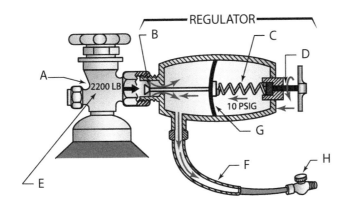

A _____

B _____

C _____

D _____

E _____

F _____

G _____

H _____

Figure 17-1

Instructions: In the space provided, identify the items shown in the illustration.

Name: _____ Date: _____

Class: _____ Instructor: _____ Score: _____

A. _____

B. _____

C. _____

D. _____

E. _____

Figures 17-4 and 17-5

Instructions: In the space provided, identify the items shown in the illustrations.

Name: _____ Date: _____

Class: _____ Instructor: _____ Score: _____

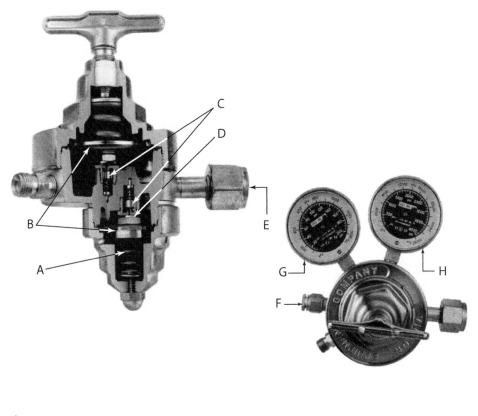

A_____

B_____

C_____

D_____

E_____

F_____

G_____

H_____

Figures 17-6 and 17-21

Instructions: In the space provided, identify the items shown in the illustrations.

Name:_____ Date:_____

Class:_____ Instructor:_____ Score:_____

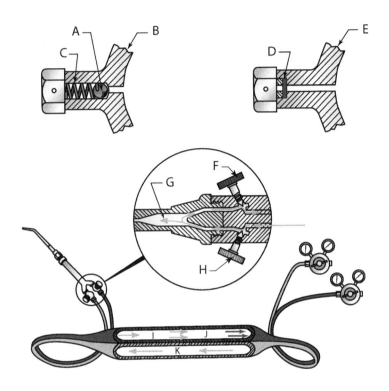

A _____

B _____

C _____

D _____

E _____

F _____

G _____

H _____

I _____

J _____

K _____

Figures 17-6 and 17-21

Instructions: In the space provided, identify the items shown in the illustrations.

Name: _____ Date: _____

Class: _____ Instructor: _____ Score: _____

Figures 17-36 and 17-43

Instructions: In the space provided, identify the items shown in the illustrations.

Name: _____ Date: _____

Class: _____ Instructor: _____ Score: _____

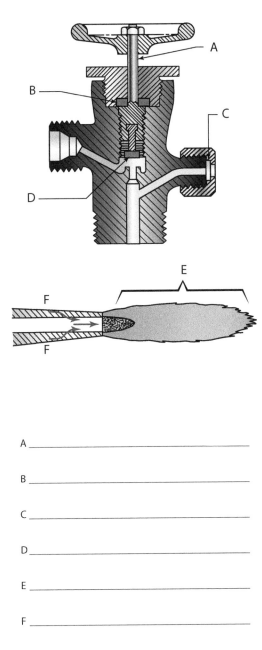

A _____

B _____

C _____

D _____

E _____

F _____

334

Figure Identification Exercises

Figures 17-36 and 17-43

Instructions: In the space provided, identify the items shown in the illus. below.

Name _____ Date _____

Class _____ Instructor _____ Score _____

Figures 18-4 and 18-9

Instructions: In the space provided, identify the items shown in the illustrations.

Name: _____ Date: _____

Class: _____ Instructor: _____ Score: _____

A _____   G _____

B _____   H _____

C _____   I _____

D _____   J _____

E _____   K _____

F _____

Figures 18-4 and 18-9

Instructions: In the space provided, identify the items shown in the illustrations.

Name _____ Date _____

Class _____ Instructor _____ Score _____

A _____   G _____

B _____   H _____

C _____   I _____

D _____   J _____

E _____   K _____

F _____

Figures 18-13, 18-18, and 18-21

Instructions: In the space provided, identify the items shown in the illustrations.

Name: _____ Date: _____

Class: _____ Instructor: _____ Score: _____

A _____

B _____

C _____

D _____

E _____

F _____

G _____

H _____

I _____

Figures 18-25, 18-47, and 18-48

Instructions: In the space provided, identify the items shown in the illustrations.

Name: _____ Date: _____

Class: _____ Instructor: _____ Score: _____

A _____

B _____

C _____

D _____

E _____

F _____

G _____

H _____

Figures 18-25, 18-47, and 18-48

Instructions: In the space provided, identify the items shown in the illustrations.

Name _____ Date _____

Class _____ Instructor _____ Score _____

Figures 19-13 and 19-24

Instructions: In the space provided, identify the items shown in the illustrations.

Name: _____ Date: _____

Class: _____ Instructor: _____ Score: _____

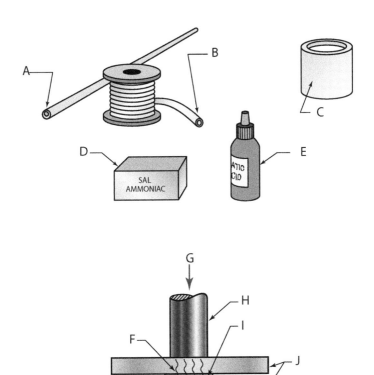

A _____   G _____

B _____   H _____

C _____   I _____

D _____   J _____

E _____   K _____

F _____   L _____

Figures 19-13 and 19-24

Instructions: In the space provided, identify the items shown in the illustration.

Name: _____ Date: _____

Class: _____ Score: _____ Instructor: _____

Figure 19-18

Instructions: In the space provided, identify the items shown in the illustrations.

Name:_____ Date:_____

Class:_____ Instructor:_____ Score:_____

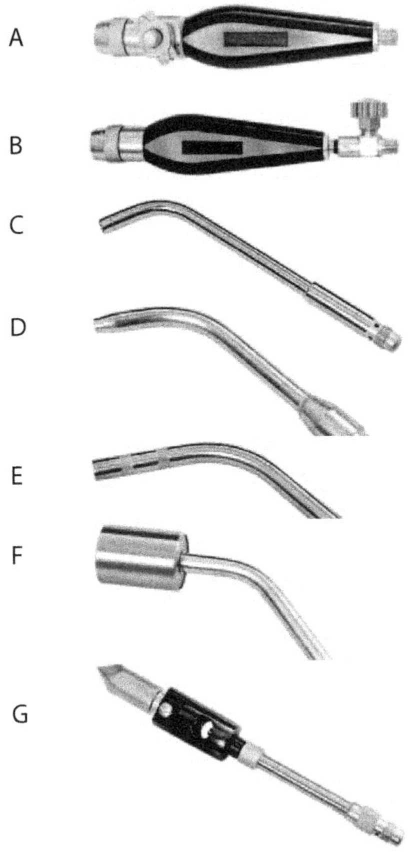

A

B

C

D

E

F

G

A _____

B _____

C _____

D _____

E _____

F _____

G _____

Figure 19-18

Instructions: In the space provided, identify the items shown in the illustration.

A

B

C

D

Figures 19-20 and 19-23

Instructions: In the space provided, identify the items shown in the illustrations.

Name: _____ Date: _____

Class: _____ Instructor: _____ Score: _____

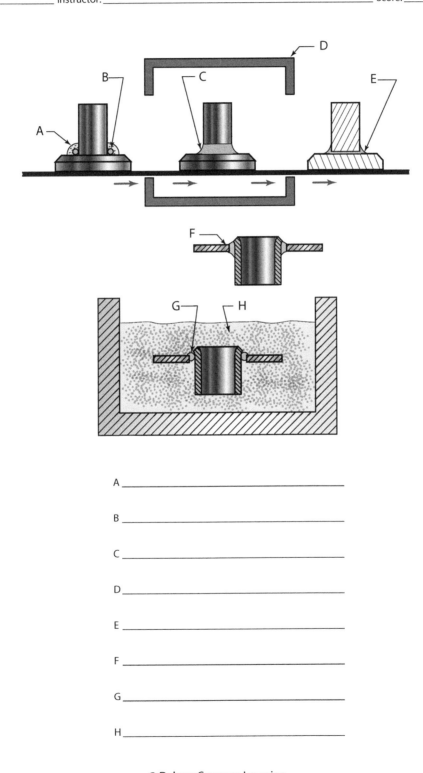

A _____

B _____

C _____

D _____

E _____

F _____

G _____

H _____

Instructions: In the space provided, identify the signs shown in the illustration.

Name _____ Date _____

Class _____ Instructor _____

Figures 20-2 and 20-13

Instructions: In the space provided, identify the items shown in the illustrations.

Name:_____ Date:_____

Class:_____ Instructor:_____ Score:_____

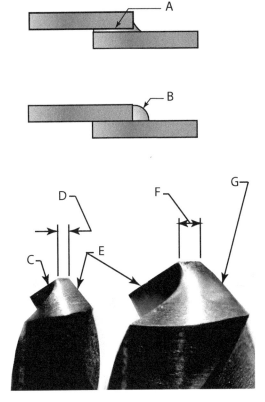

A _____

B _____

C _____

D _____

E _____

F _____

G _____

Figures 20-2 and 20-13

Instructions: In the space provided, identify the items shown in the illustrations.

Name: _____ Date: _____

Class: _____ Instructor: _____ Score: _____

Figure 20-17

Instructions: In the space provided, identify the items shown in the illustrations.

Name: _____ Date: _____

Class: _____ Instructor: _____ Score: _____

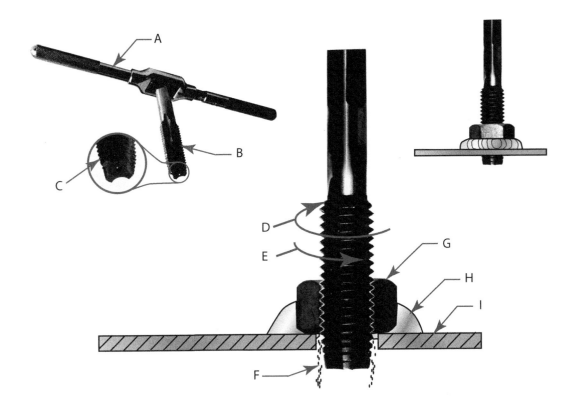

A _____

B _____

C _____

D _____

E _____

F _____

G _____

H _____

I _____

Figure 20-17

Instructions: In the space provided, identify the items shown in the illustration.

Name _____ Date _____

Class _____ Instructor _____

Figures 21-2, 21-3A, 21-3C, 21-6, and 21-10

Instructions: In the space provided, identify the items shown in the illustrations.

Name:_____ Date:_____

Class:_____ Instructor:_____ Score:_____

A _____

B _____

C _____

D _____

E _____

F _____

G _____

H _____

Figures 21-2, 21-3A, 21-3C, 21-6, and 21-10

Instructions: In the space provided, identify the items shown in the illustrations.

Name: _____ Date: _____

Class: _____ Instructor: _____ Score: _____

A. _____

B. _____

C. _____

D. _____

E. _____

F. _____

G. _____

H. _____

Figure 21-4

Instructions: In the space provided, identify the items shown in the illustration.

Name: _____ Date: _____

Class: _____ Instructor: _____ Score: _____

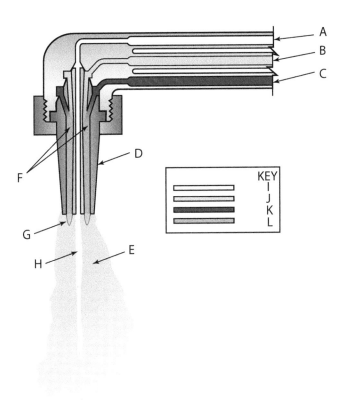

A _____   H _____

B _____   I _____

C _____   J _____

D _____   K _____

E _____   L _____

F _____

G _____

Figure 21-4

Instructions: In the space provided, identify the items shown in this illustration.

Name: _____ Date: _____

Class: _____ Instructor: _____

A _____

B _____

C _____

D _____

E _____

F _____

G _____

Figures 21-18 and 21-23

Instructions: In the space provided, identify the items shown in the illustrations.

Name:_____ Date:_____

Class:_____ Instructor:_____ Score:_____

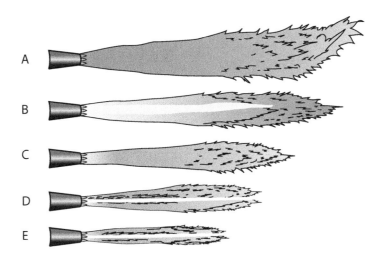

F

A _____

B _____

C _____

D _____

E _____

F _____

Figures 21-18 and 21-23

Instructions: In the space provided, identify the items shown in the illustrations.

Name: _____ Date: _____

Class: _____ Instructor: _____ Score: _____

A. _____

B. _____

C. _____

D. _____

E. _____

F. _____

G. _____

Figure 21-19

Instructions: In the space provided, identify the items shown in the illustration.

Name: _____ Date: _____

Class: _____ Instructor: _____ Score: _____

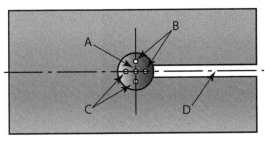

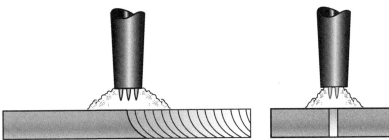

A _____

B _____

C _____

D _____

Figure 21-40

Instructions: In the space provided, identify the items shown in the illustration.

Name: _____ Date: _____

Class: _____ Instructor: _____ Score: _____

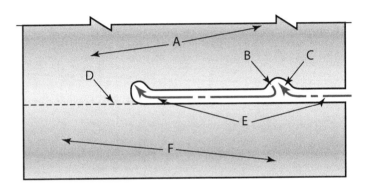

A _____

B _____

C _____

D _____

E _____

F _____

Figure 21-40

Instructions: In the space provided, identify the items shown in the illustration.

Name: _____ Date: _____

Class: _____ Instructor: _____ Score: _____

A. _____

B. _____

C. _____

D. _____

E. _____

F. _____

Figure 22-4

Name: _____ Date: _____

Class: _____ Instructor: _____ Score: _____

TEMPERATURES

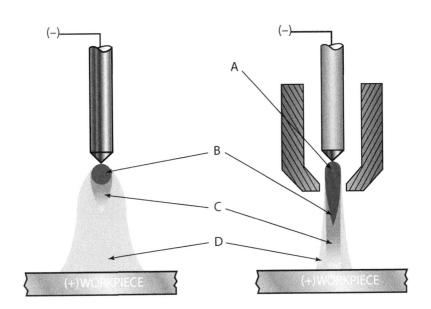

A _____

B _____

C _____

D _____

Figure 22-4

Instructions: In the space provided, identify the items shown in the illustration.

Name _____ Date _____

Class _____ Instructor _____ Score _____

TEMPERATURE

A. _____

B. _____

C. _____

D. _____

Figure 22-5

Instructions: In the space provided, identify the items shown in the illustration.

Name: _____ Date: _____

Class: _____ Instructor: _____ Score: _____

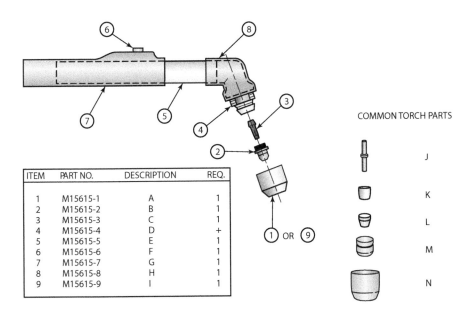

COMMON TORCH PARTS

ITEM	PART NO.	DESCRIPTION	REQ.
1	M15615-1	A	1
2	M15615-2	B	1
3	M15615-3	C	1
4	M15615-4	D	+
5	M15615-5	E	1
6	M15615-6	F	1
7	M15615-7	G	1
8	M15615-8	H	1
9	M15615-9	I	1

A _____   H _____

B _____   I _____

C _____   J _____

D _____   K _____

E _____   L _____

F _____   M _____

G _____   N _____

Figure 22-5

Instructions: In the space provided, identify the items shown in the illustration.

Name: _____ Date: _____

Class: _____ Instructor: _____

COMMON TORQUE UNITS

Figure 22-8

Instructions: In the space provided, identify the items shown in the illustration.

Name: _____ Date: _____

Class: _____ Instructor: _____ Score: _____

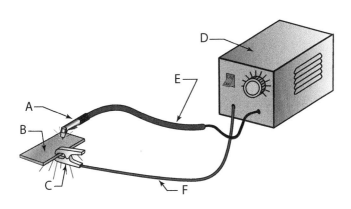

A _____

B _____

C _____

D _____

E _____

F _____

Figure 22-3

Instructions: In the space provided, identify the items shown in the illustration.

Name _____ Date _____

Class _____ Instructor _____ Score _____

Figure 22-15

Name: _____ Date: _____

Class: _____ Instructor: _____ Score: _____

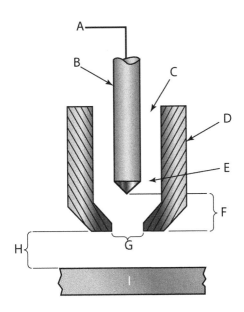

A _____

B _____

C _____

D _____

E _____

F _____

G _____

H _____

I _____

Figure 22-15

Instructions: In the space provided, identify the items shown in the illustration.

Name: _____ Date: _____

Class: _____ Instructor: _____ Score: _____

A. _____

B. _____

C. _____

D. _____

E. _____

F. _____

G. _____

H. _____

I. _____

Figure 22-19

Instructions: In the space provided, identify the items shown in the illustration.

Name: _____ Date: _____

Class: _____ Instructor: _____ Score: _____

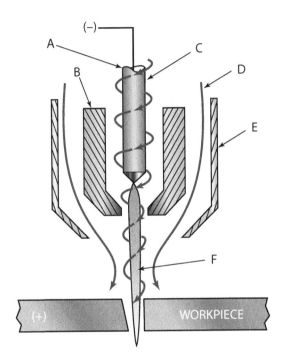

A _____

B _____

C _____

D _____

E _____

F _____

Figure 22-19

Instructions: In the space provided, identify the items shown in the illustration.

Name _____ Date _____

Class _____ Instructor _____ Score _____

A. _____

B. _____

_____

_____

F. _____

Figure 23-6

Instructions: In the space provided, identify the items shown in the illustration.

Name:_____ Date:_____

Class:_____ Instructor:_____ Score:_____

Energy Monitor

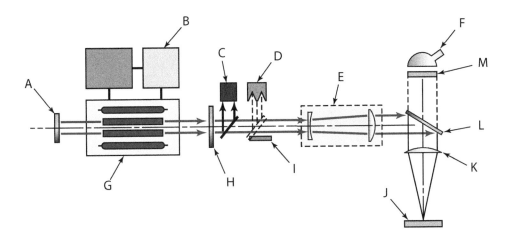

A _____   H_____

B _____   I _____

C _____   J _____

D _____   K _____

E _____   L _____

F _____   M_____

G _____

Figure 23-6

Instructions: In the space provided, identify the items shown in the illustration.

Name _____ Date _____

Class _____ Instructor _____ Score _____

Energy Monitor

A. _____

B. _____

C. _____

D. _____

E. _____

M. _____

Figure 23-13

Instructions: In the space provided, identify the items shown in the illustration.

Name: _____ Date: _____

Class: _____ Instructor: _____ Score: _____

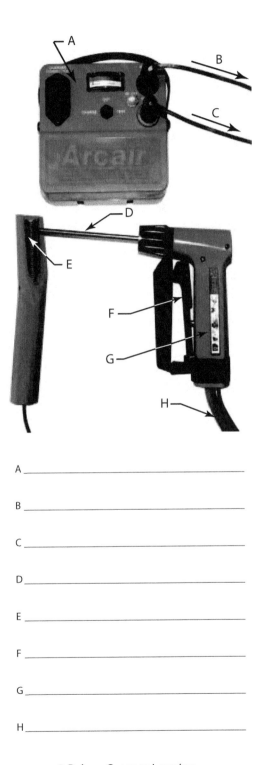

A _____

B _____

C _____

D _____

E _____

F _____

G _____

H _____

Figure 23-13

Instructions: In the space provided, identify the items shown in the illustration.

Name _____ Date _____

Class _____ Instructor: _____ Score _____

Figure 23-20

Instructions: In the space provided, identify the items shown in the illustration.

Name: _____ Date: _____

Class: _____ Instructor: _____ Score: _____

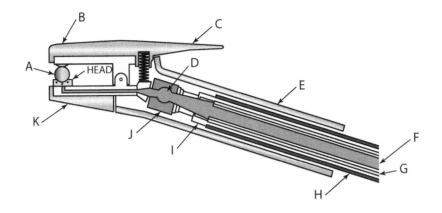

A _____

B _____

C _____

D _____

E _____

F _____

G _____

H _____

I _____

J _____

K _____

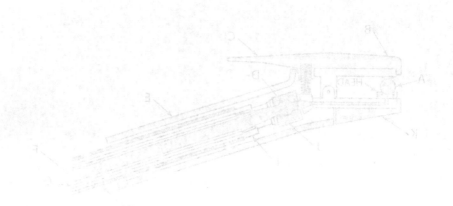

Figure 23-20

Instructions: In the space provided, identify the items shown in the illustration.

Name: _____   Date _____

Class _____   Instructor: _____   Score _____

A. _____

B. _____

C. _____

D. _____

E. _____

F. _____

G. _____

H. _____

I. _____

J. _____

K. _____

Figure 23-22

Instructions: In the space provided, identify the items shown in the illustration.

Name: _____ Date: _____

Class: _____ Instructor: _____ Score: _____

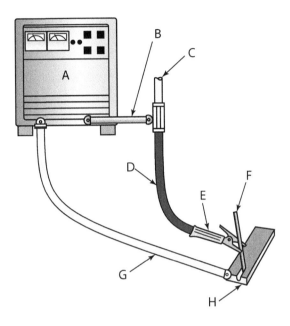

A _____

B _____

C _____

D _____

E _____

F _____

G _____

H _____

Figure 23-22

Instructions: In the space provided, identify the items shown in the illustration.

Name: _____ Date: _____

Class: _____ Instructor: _____ Score: _____

Figure 24-4

Name:_____ Date:_____

Class:_____ Instructor:_____ Score:_____

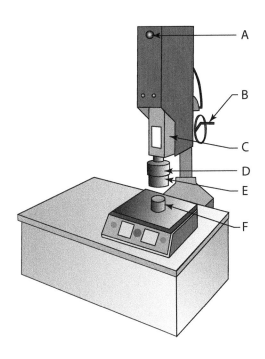

A _____

B _____

C _____

D _____

E _____

F _____

Figure 24-4

Instructions: In the space provided, identify the items shown in the illustration.

Name: _____ Date: _____

Class: _____ Instructor: _____ Score: _____

A. _____

B. _____

C. _____

D. _____

E. _____

F. _____

Figure 24-5

Instructions: In the space provided, identify the items shown in the illustration.

Name: _____ Date: _____

Class: _____ Instructor: _____ Score: _____

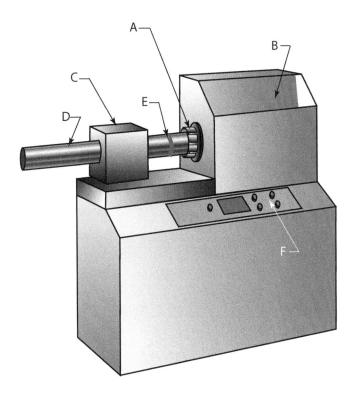

A _____

B _____

C _____

D _____

E _____

F _____

Figure 24-5

Instructions: In the space provided, identify the items shown in the illustration.

Name _____ Date _____

Class _____ Instructor _____ Score _____

A. _____

B. _____

C. _____

D. _____

E. _____

Figure 24-8

Instructions: In the space provided, identify the items shown in the illustration.

Name: _____ Date: _____

Class: _____ Instructor: _____ Score: _____

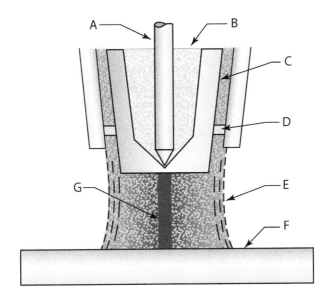

A _____

B _____

C _____

D _____

E _____

F _____

G _____

Figure 24-8

Instructions: In the space provided, identify the items shown in the illustration.

Name: _____ Date: _____

Class: _____ Instructor: _____ Score: _____

A. _____

B. _____

C. _____

D. _____

E. _____

F. _____

G. _____

Figure 24-12

Instructions: In the space provided, identify the items shown in the illustration.

Name: _____ Date: _____

Class: _____ Instructor: _____ Score: _____

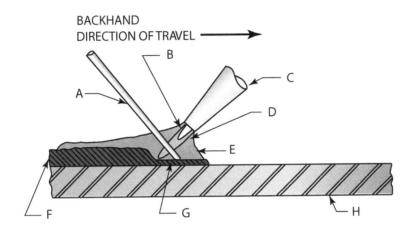

BACKHAND
DIRECTION OF TRAVEL →

A _____

B _____

C _____

D _____

E _____

F _____

G _____

H _____

Figure 24-12

Instructions: In the space provided, identify the items shown in the illustration.

Name _____ Date _____

Class _____ Score _____ Instructor _____

BACKHAND
DIRECTION OF TRAVEL

Figure 24-16

Instructions: In the space provided, identify the items shown in the illustration.

Name: _____ Date: _____

Class: _____ Instructor: _____ Score: _____

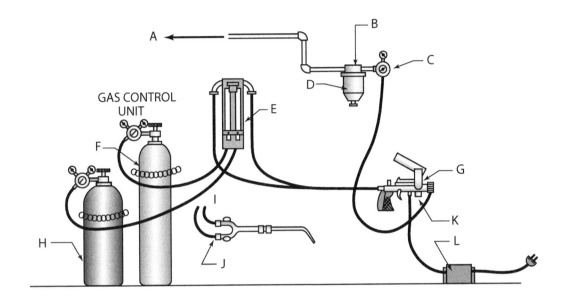

A _____

B _____

C _____

D _____

E _____

F _____

G _____

H _____

I _____

J _____

K _____

L _____

Figure 24-16

Instructions: In the space provided, identify the items shown in the illustration.

Name: _____ Date: _____

Class: _____ Instructor: _____ Score: _____

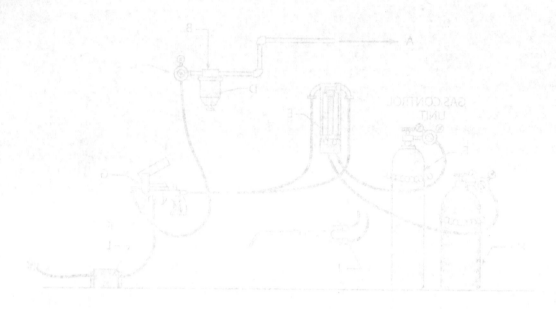

GAS CONTROL UNIT

A. _____

B. _____

C. _____

D. _____

E. _____

F. _____

G. _____

H. _____

I. _____

J. _____

K. _____

L. _____

Figures 25-1 and 25-4

Instructions: In the space provided, identify the items shown in the illustrations.

Name: _____ Date: _____

Class: _____ Instructor: _____ Score: _____

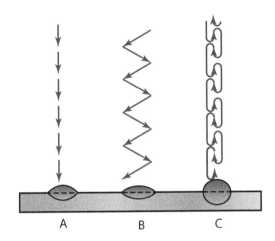

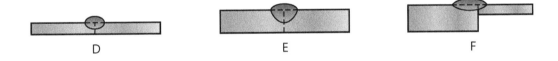

A _____

B _____

C _____

D _____

E _____

F _____

Figure 25-8

Instructions: In the space provided, identify the items shown in the illustration.

Name:_____ Date:_____

Class:_____ Instructor:_____ Score:_____

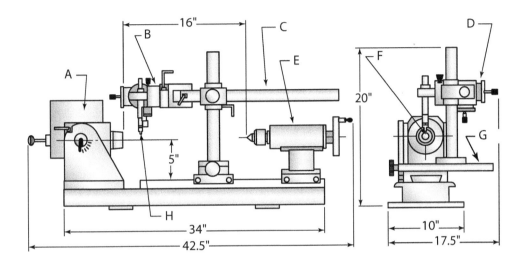

A_____

B_____

C_____

D_____

E_____

F_____

G_____

H_____

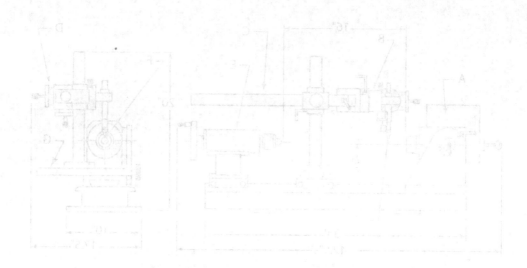

Figures 26-5, 26-14, and 26-15

Instructions: In the space provided, identify the items shown in the illustrations.

Name: _____ Date: _____

Class: _____ Instructor: _____ Score: _____

E 6 0 1 – 2

```
A
B
C
D
```

E R XX S – X

```
E
F
G
H
I
```

E X X T – X

```
J
K
L
M
N
```

A _____   H _____

B _____   I _____

C _____   J _____

D _____   K _____

E _____   L _____

F _____   M _____

G _____   N _____

Figures 27-1, 27-3, and 27-5

Instructions: In the space provided, identify the items shown in the illustrations.

Name:_____ Date:_____

Class:_____ Instructor:_____ Score:_____

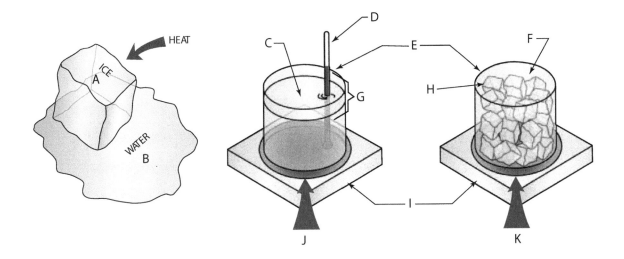

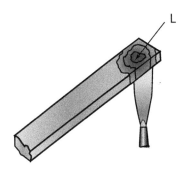

A _____   G _____

B _____   H _____

C _____   I _____

D _____   J _____

E _____   K _____

F _____   L _____

Figures 27-1, 27-3, and 27-5

Instructions: In the space provided, identify the items shown in the illustrations.

Name _____ Date _____

Class _____ Instructor _____ Score _____

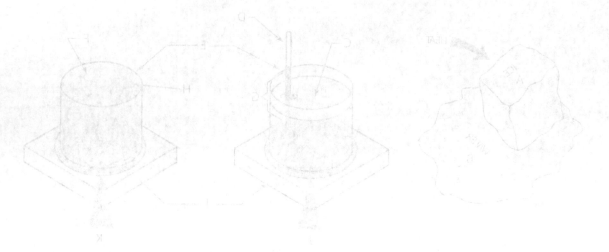

A. _____

B. _____

C. _____

D. _____

E. _____

F. _____

Figures 27-6–27-8

Instructions: In the space provided, identify the items shown in the illustrations.

Name: _____ Date: _____

Class: _____ Instructor: _____ Score: _____

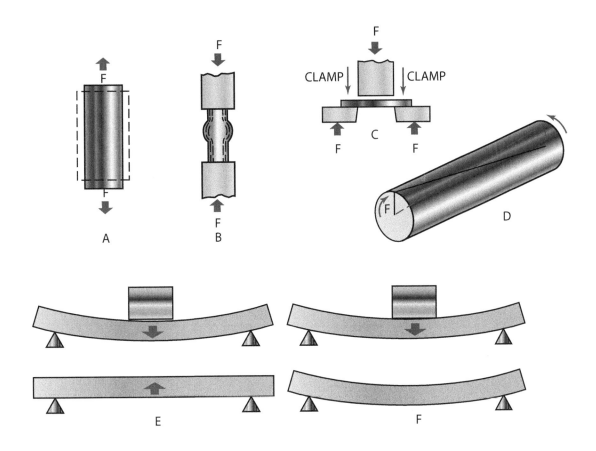

A _____

B _____

C _____

D _____

E _____

F _____

Figures 27-18, 27-19, 27-20, and 27-29

Instructions: In the space provided, identify the items shown in the illustrations.

Name: _____ Date: _____

Class: _____ Instructor: _____ Score: _____

A _____

B _____

C _____

D _____

E _____

F _____

G _____

H _____

I _____

J _____

Figures 27-18, 27-19, 27-20, and 27-29

Instructions. In the space provided, identify the items shown in the illustration.

Name: _____ Date: _____

Class: _____ Instructor: _____ Score: _____

A. _____

B. _____

C. _____

D. _____

E. _____

F. _____

G. _____

H. _____

Figure 27-24

Instructions: In the space provided, identify the items shown in the illustration.

Name: _____ Date: _____

Class: _____ Instructor: _____ Score: _____

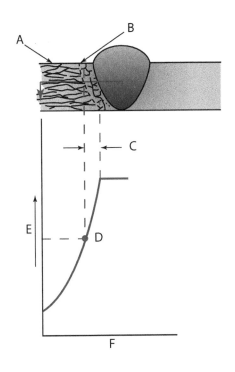

A _____

B _____

C _____

D _____

E _____

F _____

Figure 27-26

Instructions: In the space provided, identify the items shown in the illustration.

Name: _____ Date: _____

Class: _____ Instructor: _____ Score: _____

Figure 28.T2

Instructions: In the space provided, identify the items.

Name:_____ Date:_____

Class:_____ Instructor:_____ Score:_____

_____	Manganese 1.75
_____	Nickel 3.50
_____	Nickel 5.00
_____	Nickel 1.25; chromium 0.65
_____	Nickel 3.50; chromium 1.55; electric furnace
_____	Molybdenum 0.25
_____	Chromium 0.50 or 0.95; molybdenum 0.12 or 0.20
_____	Nickel 1.80; chromium 0.50 or 0.80; molybdenum 0.25
_____	Same as above, produced in basic electric furnace
_____	Manganese 0.80; molybdenum 0.40
_____	Nickel 1.85; molybdenum 0.25
_____	Nickel 1.05; chromium 0.45; molybdenum 0.20 or 0.35
_____	Chromium 0.28 or 0.40
_____	Chromium 0.80, 0.88, 0.93, 0.95, or 1.00
_____	High carbon; high chromium; electric furnace bearing steel
_____	Carbon 1.00; chromium 0.50
_____	Carbon 1.00; chromium 1.00
_____	Carbon 1.00; chromium 1.45
_____	Chromium 0.60, 0.80, or 0.95; vanadium 0.12, or 0.10, or 0.15 minimum
_____	Carbon 0.40; chromium 1.60; molybdenum 0.35; aluminum 1.15
_____	Nickel 0.30; chromium 0.40; molybdenum 0.12
_____	Nickel 0.55; chromium 0.50; molybdenum 0.20
_____	Nickel 0.55; chromium 0.50; molybdenum 0.25
_____	Nickel 0.55; chromium 0.50; molybdenum 0.35
_____	Manganese 0.85; silicon 2.00; 9262-chromium 0.25 to 0.40
_____	Nickel 3.25; chromium 1.20; molybdenum 0.12
_____	Nickel 1.00; chromium 0.80; molybdenum 0.25
_____	Boron
_____	Chromium 0.50 or 0.28; boron
_____	Chromium 0.80; boron
_____	Nickel 0.33; chromium 0.45; molybdenum 0.12; boron
_____	Nickel 0.55; chromium 0.50; molybdenum 0.20; boron
_____	Nickel 0.45; chromium 0.40; molybdenum 0.12; boron

Note: The elements in this table are expressed in percent.

*Consult current AISI and SAE publications for the latest revisions.

**Nonstandard steel.

Figure 29-T12

Instructions: In the space provided, identify the items shown in the illustration.

Name: _____ Date: _____

Class: _____ Instructor: _____ Score: _____

Type of Material	
P-1	A
P-3	B
P-4	C
P-5	D
P-6	E
P-7	F
P-8	G
P-9	H
P-10	I
P-21	J
P-31	K
P-41	L

A _____

B _____

C _____

D _____

E _____

F _____

G _____

H _____

I _____

J _____

K _____

L _____

Figure 29-T14

Instructions: In the space provided, identify the items shown in the illustration.

Name: _____ Date: _____

Class: _____ Instructor: _____ Score: _____

A-Number	Metal and Process(es)
A5.10	A
A5.3	B
A5.8	C
A5.1	D
A5.20	E
A5.17	F
A5.18	G
A5.2	H
A5.5	I
A5.23	J
A5.28	K
A5.29	L

A _____

B _____

C _____

D _____

E _____

F _____

G _____

H _____

I _____

J _____

K _____

L _____

Figure 29-T15

Instructions: In the space provided, identify the items shown in the illustration.

Name: _____ Date: _____

Class: _____ Instructor: _____ Score: _____

Group Designation	Metal Types	AWS Electrode Classification
F1	Carbon steel	A
F2	Carbon steel	B
F3	Carbon steel	C
F4	Carbon steel	D
F5	Stainless steel	E
F6	Stainless steel	F
F22	Aluminum	G

A _____

B _____

C _____

D _____

E _____

F _____

G _____

Figure 29-15

Instructions: In the space provided, identify the items shown in the illustration.

Name _____ Date _____

Class _____ Instructor _____ Score _____

Color Designation	Metal Type	AISI Number Classification
F1	Cast in steel	
	Carbon steel	
	Carbon steel	

Figures 30-11, 30-12, 30-15, and 30-17

Instructions: In the space provided, identify the items shown in the illustrations.

Name: _____ Date: _____

Class: _____ Instructor: _____ Score: _____

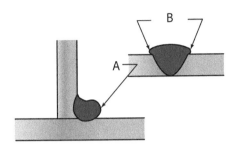

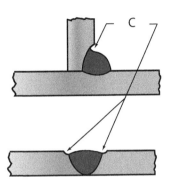

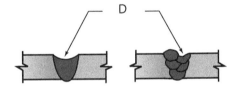

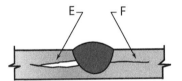

A _____

B _____

C _____

D _____

E _____

F _____

Figures 30-31 and 30-40

Instructions: In the space provided, identify the items shown in the illustrations.

Name:_____ Date:_____

Class:_____ Instructor:_____ Score:_____

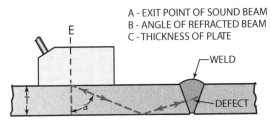

A - EXIT POINT OF SOUND BEAM
B - ANGLE OF REFRACTED BEAM
C - THICKNESS OF PLATE

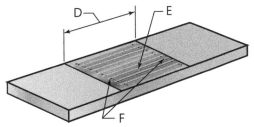

A _____

B _____

C _____

D _____

E _____

F _____

Figure 30-37

Instructions: In the space provided, identify the items shown in the illustration.

Name:_____ Date:_____

Class:_____ Instructor:_____ Score:_____

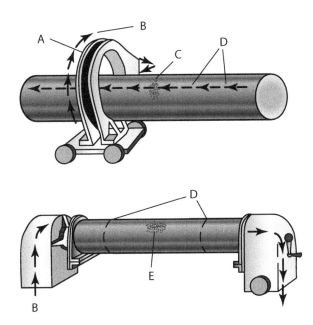

A _____

B _____

C _____

D _____

E _____